JUPITER

Kosmos

A series exploring our expanding knowledge of the cosmos through science and technology and investigating historical, contemporary and future developments as well as providing guidance for all those interested in astronomy.

Series Editor: Peter Morris

Already published:

Asteroids Clifford J. Cunningham
The Greatest Adventure Colin Burgess
Jupiter William Sheehan and Thomas Hockey
Mars Stephen James O'Meara
Mercury William Sheehan
The Moon Bill Leatherbarrow
Saturn William Sheehan
Soviets in Space Colin Burgess
The Sun Leon Golub and Jay M. Pasachoff
Uranus and Neptune Carolyn Kennett
Venus William Sheehan and Sanjay Shridhar Limaye

Jupiter

William Sheehan
and Thomas Hockey

REAKTION BOOKS

To John Rogers (W. S.)
To Reta Beebe (T. H.)

Published by Reaktion Books Ltd
Unit 32, Waterside
44–48 Wharf Road
London N1 7UX, UK
www.reaktionbooks.co.uk

First published 2018
First published in paperback 2023

Printed and bound in India by Replika Press Pvt. Ltd.

A catalogue record for this book is available from the British Library

ISBN 978 1 78914 705 6

CONTENTS

One of the best pictures of Jupiter obtained from the Earth, by the doyen of amateur planetary imagers, Damian Peach. Peach used an ASI 174-mm camera on a 1-m Cassegrain located near Cerro Pachon, Chile, that he operated remotely from his home in Hamble, Hampshire, England.

PROLOGUE

To the naked eye, Venus is brighter, while Mars, which appears like a red-hot coal round the times of its every-other-year oppositions, is more dramatic. But Jupiter, even to the naked eye, is still the grandest of all the planets. It shines with a majestic steady mien and is conspicuous for about ten of the thirteen months that elapse between its successive conjunctions with the Sun. (For about three months of each of these thirteen, it is lost in the solar glare as it passes behind the Sun.) It has inspired more poetic utterances than any other planet, with the single exception of Venus, the beautiful Morning and Evening Star. William Wordsworth beautifully invokes Jupiter in the *Prelude*, Book IV (1850):

> A pensive feeling spread far and wide:
> The trees, the mountains shared it, and the brooks.
> The stars of heaven, now seen in their old haunts –
> White Sirius glittering o'er the southern crags,
> Orion with his belt, and those fair seven,
> Acquaintances of every little child,
> And Jupiter, my own beloved star.

Walt Whitman also invokes its majestic calming presence in 'On the Beach at Night', from *Leaves of Grass* (1881–2):

On the beach at night,
Stands a child with her father,
Watching the east, the autumn sky.
Up through the darkness,
While ravening clouds, the buried
Clouds, in black masses spreading,
Lower sullen and fast athwart and down the sky,
Amid a transparent clear belt of ether
Yet left in the east,
Ascends large and calm the lord-star Jupiter,
And nigh at hand, only a very little above,
Swim the delicate stars the Pleiades.

The planet has evoked such feelings as far back as we have records, for it was no less than the 'beloved star' of the first serious observers of the planets, the ancient Sumerians and Babylonians. Three thousand years ago they admired its beauty and calm grandeur as it stood in the mid-heavens, regal and untroubled above the gathering clouds in the east. And they too named it 'lord-star'.

The Sumerians and Babylonians did not regard the planets as gods as many other early peoples, including the Greeks, did. Instead they saw them as manifestations, interpreters; they were the 'stars of the great gods who ruled the world'. Their wanderings among the background stars, reversals of direction and conjunctions – the other planets and prominent stars were viewed as omens providing a cryptic commentary on terrestrial affairs. Many of these omens were collected in the so-called 'Enuma Anu Enlil' tablets, a series of seventy tablets containing thousands of omens dating back to the second millennium BCE. The name 'Enuma Anu Enlil' means 'when Anu and Enlil . . .', the opening words of the first tablet. Anu and Enlil were Sumerian gods while the chief god of the Babylonian pantheon was Marduk (associated with Jupiter). The following is typical of the omens: 'If Marduk [rises] in the path of the [god

The famous ziggurat at Ur in Mesopotamia. In the background shines the Star of Marduk, the chief Babylonian god. Painting by Julian Baum, 2009.

Enlil's] stars, the king of Akkad will become strong and [overthrow] his enemies in all lands in battle.'[1]

Much later, in China in the fourth century BCE, the astrologer Gan De compiled *Suixing Jing* (*A Canon of the Planet Jupiter*), now lost. The Chinese name for the planet was Muxing. As Marduk had been for the Babylonians, Muxing was for the Chinese: the head of all the host of heaven.

Gan De had a close associate, Shi Shen, and together they produced a star catalogue, some two centuries before the famous star catalogue of the Greek astronomer Hipparchus. (Earlier Babylonian star catalogues exist but the authors are anonymous.) Though the Chinese were especially interested in Jupiter, they made accurate observations of the other planets as well, and worked out their sidereal and synodic periods. These periods are still fundamental in astronomy and are needed for planning observations of the planets.

The synodic period is the time a planet takes to return to the same position on the celestial sphere relative to the Sun. The word comes from the Greek *synod*, which means a meeting or gathering. This term is well known in connection with ecclesiastical assemblies and councils, but its use in astronomy refers simply to the successive meetings of a planet and the Sun. The sidereal period is the time it takes a planet to return to the same location with respect to the stars.

A table of synodic and sidereal periods for the five planets visible to the naked eye follows. The periods given are modern ones. Expressed in terms of one another, the synodic period = product of the sidereal periods of two planets ÷ difference of the sidereal periods. In the table, the synodic periods are given relative to the Earth.

Marduk, the Babylonian 'Jupiter'.

Since Jupiter's synodic period is about a year and a month, each time it appears opposite the Sun in the sky (a configuration known as opposition, in which it rises when the Sun sets and sets when the Sun rises), it appears to have moved one zodiacal constellation over from the last opposition. After twelve years, the cycle repeats. This is evident from the following table, which lists opposition dates with the constellations in which Jupiter appears at the time of opposition.

The sidereal and synodic periods were well known to the Babylonians, who passed them on to the Greeks in the fourth century BCE, following the conquest of Babylon by Alexander the Great. Henceforth they were common knowledge among scholars throughout the Mediterranean basin. However, since there was no cross-pollination between the astronomy of the Mediterranean world and that of the Far East at the time, the ancient Chinese had to discover them independently. Because of Jupiter's evident importance, its 11.86-year sidereal period actually defined the constellations of the Chinese zodiac

TABLE I: *Planetary Sidereal and Synodic Periods*

Planet	Sidereal Period	Synodic Period
Mercury	88.97 days (0.2409 yrs)	115.88 days (0.317 yrs)
Venus	225 days (0.6152 yrs)	583.9 days (1.599 yrs)
Earth	365.25636 solar days	(1 year)
Mars	1.881 years	779.9 days (2.135 yrs)
Jupiter	11.86 years	398.9 days (1.092 yrs)
Saturn	29.46 years	378.1 days (1.035 yrs)

TABLE II: *Oppositions of Jupiter, 2017–29*

Date	Constellation
7 April 2017	Libra
9 May 2018	Scorpio
10 June 2019	Sagittarius
14 July 2020	Capricorn
20 August 2021	Aquarius
26 September 2022	Aries
3 November 2023	Taurus
7 December 2024	Gemini
10 January 2026	Cancer
11 February 2027	Leo
12 March 2028	Virgo
12 April 2029	Libra

(*Shengxiao*). The Chinese constellations correspond only roughly with those used by the Greeks, but they are more consistent – theirs is a true *zoidiakos*, or circle of animals, whereas only half the Greek zodiacal constellations are so (Pisces the Fish, Aries the Ram, Taurus the Bull, Cancer the Crab, Leo the Lion and Scorpio the Scorpion). The constellations of the Chinese zodiac are the ox, the tiger, the rabbit, the dragon, the snake, the horse, the goat, the monkey, the rooster, the dog and the pig, which will be familiar to anyone who has ever eaten in a Chinese restaurant and studied a place mat defining birth years as the 'year of the ox', the 'year of the goat' and so on. Few who have done so realize the connection with the planet Jupiter.

One ancient symbol for Jupiter, ♃, is still in use today. It is meant to represent a lightning bolt flung by the god Jupiter.

Majestic Jupiter

While not as brilliant as Venus, Jupiter is steadier-going and less fickle. Its position as a superior planet, orbiting further from the Sun than the Earth, means that its lustre does not vary as greatly as that of Venus, which, as an inferior planet orbiting closer to the Sun than the Earth, often 'hides' from our eyes in conjunction when it is lined up on either the near side (inferior conjunction) or far side (superior conjunction) of the Sun. In contrast, moreover, to Venus, which never ventures far from the Sun, and so can be seen only for a few hours before sunrise or after sunset, Jupiter can be seen opposite the Sun in the sky (when it is in opposition). It then shines brightly from sunset to sunrise, and can be seen lording it overhead at midnight.

Even remote Saturn changes more in brightness than does Jupiter. All the planets shine by reflecting sunlight, and because of the long-term variation in the tilt of Saturn's rings with respect to the Earth, Saturn's light-reflecting surface (ball plus rings) alternately increases and decreases; so, accordingly, does its apparent brightness.

The steadiness and constancy of Jupiter's light resembles that of the Sun itself. The Maya, a New World civilization that made great accomplishments in naked-eye astronomy, noticed this, and went so far as to describe it as the 'Night Sun'. The fact that it is steady and untwinkling is owed to the fact that, rather than being a stellar point source, Jupiter poses a small planetary disc. Thus its light is not as affected by currents in the air.

The Roman god Jupiter, king of the gods of Mount Olympus.

It is also among the brightest objects in the night sky. It can be three times brighter than Sirius, the brightest star in the sky, and on a dark, moonless night, it is easily able to cast shadows. It can just be glimpsed in broad daylight with the naked eye by keen-sighted individuals.

In terms of the quantitative magnitude scale, Jupiter reaches an apparent magnitude of $m = -2.6$. (The more negative the number, the brighter the celestial object.) In comparison, the brightest star, Sirius, is $m = -1.4$, and the Full Moon is $m = -12.5$. Jupiter is the fourth brightest of the objects that are usually visible in the sky, after the Sun, Moon and Venus. Mars, at extremely favourable approaches – as in 2003 and in 2018 – can also just outshine Jupiter.

The ancients who called it Marduk, Muxing and Jupiter, the last for the king of the gods of Mount Olympus, intuited that there was something noble and majestic about this planet. They are proven to have been spot on. The apparent size of Jupiter's disc is larger than that of the other planets, except Venus's near inferior conjunction. Its large apparent size and steady performance in our night skies is due to the fact that it really is immense: it is the largest planet in the solar system, and so vast that 1,300 globes the size of the Earth could be fitted inside with room to spare. Majestic indeed!

Because Jupiter and the other planets all travel on similar paths, circling through the zodiacal constellations as viewed from the Earth, one planet may *appear* to catch up with another and appear side by side with it in the night sky. (Even though they appear to be close, remember that the planets are actually at terrific distances from one another.) Such an event is called a conjunction, and Jupiter appearing near another planet is a beautiful but fairly infrequent sight.

Even more infrequently, three planets may appear near one another in a triple conjunction, which is rather analogous to a slot machine coming up all oranges. When the trio of bright planets, Jupiter, Saturn and Mars, all gather together in the sky, as they appear along the same line of sight from the Earth, it is called a

Jupiter rotates, and as it does it brings fascinating details into view. This series of images was taken on the night of 12–13 March 2017 by the skilful planetary imager Leo Aerts, from Belgium, with Jupiter only 30° above the horizon. The satellite Io is the bright spot chasing the shadow. Towards the top centre in the middle image, the Red Spot, Junior, is seen, while the Great Red Spot itself follows onto the disc.

Great Conjunction (the last such occurred in 2000). Such events were long deemed of great astrological significance. The 'Star of Bethlehem' – concerning which many theories have been put forward – may have involved such a conjunction of planets. The great German astronomer Johannes Kepler believed this to have been so; he calculated that no fewer than three conjunctions of Jupiter and Saturn occurred in Pisces in May, September and December BCE, followed by a Great Conjunction of Jupiter, Saturn and Mars in the same vicinity on 6 February BCE. Such goings-on in the heavens must have seemed impressive indeed to priest-astrologers such as the Magi, as the members of the Zorastrian priestly caste of Persia (now Iran) were known.

ONE

The Jovian Planets

With many characteristics in common, the outer planets, Jupiter, Saturn, Uranus and Neptune, have sometimes been referred to as the Jovian planets after their grandest member. They are also known as the giant planets or the 'gas giants'. Worlds of a different order from the Earth and the other 'terrestrial' planets that occupy the inner solar system, they are the largest objects that we can study up close, apart from the Sun. Jupiter, in particular, is so gigantic that it has long seemed almost an embryonic star – a reputation it partly merits. Its make-up reflects this: like the Sun itself, it retains all of the gases and even some of the heat of its formation. Its bulk consists of about 90 per cent hydrogen and 10 per cent helium, reflecting the cosmic abundances of the early universe after the creation event of the Big Bang. The other Jovians have similar proportions of these primordial elements. We will devote more time later to a discussion of Jupiter's composition, and define the precise 'recipe' for making a gas giant planet.

An Almost Star is Born

Jupiter and the other planets – including the Earth – formed with the solar system itself, about 4.6 billion years ago, in a process that was long among the most inscrutable and tantalizing of the mysteries of science. After many false starts though there are still debates about some of the details, we now have a fairly good understanding of just what happened.

All peoples have origin myths. We pass over them here, and take as our starting point the Nebular Hypothesis put forward by Immanuel Kant and Pierre-Simon Laplace near the end of the eighteenth century. Kant was a German philosopher, Laplace a French mathematician, and independently of one another they tried to explain how the planets in our system came to exhibit their configuration of all moving in the same direction (in direct orbits, that is, moving anti-clockwise as seen from north of the plane of the Earth's orbit, or ecliptic) and in roughly the same plane (marked by the ecliptic). They assumed that the Sun and planets had started out as a swirling nebulous cloud that proceeded to contract gravitationally; the central mass collapsed upon itself and heated up until it could shine by its own light and heat, while the rest of the cloud flattened into a disc. Clumps of material in the disc went on to form the planets and their satellites. Note that, according to the Kant-Laplace scheme, the direction of the planets' movements was set by the initial rotatory motion of the nebula.

Immanuel Kant (1724–1804), who proposed, along with Pierre-Simon Laplace, what became known as the Kant-Laplace Nebular Hypothesis.

Although the Nebular Hypothesis as set forth by Kant and Laplace was eminently plausible, the Devil is always in the detail, and for a long time there remained an important stumbling block: angular momentum. In a nutshell, although most of the mass of the solar system resides in the slowly rotating Sun, most of the angular momentum lies in the planets. Since it seemed impossible to explain this asymmetry in terms of the Kant-Laplace scheme, a different idea came into vogue, according to which the planets formed as by-products of a rare grazing encounter of another star with the Sun in the early stages of the Sun's life as a star.

Most astronomers favoured the grazing-encounter theory at the beginning of the twentieth century. For example, in 1909 the American astronomer Percival Lowell, who established his own observatory in Flagstaff, Arizona, for the purpose of studying Mars but soon extended his programme of research to the other planets, went so far as to suggest that the spiral nebulae themselves, which were turning up by the millions in deep-sky photographs, might be budding solar systems in formation, where a dark star and a luminous star were involved in such an encounter. In that case there might be untold millions of solar systems scattered across space. Lowell wrote:

Pierre-Simon Laplace (1749–1827), in old age.

> Suppose, now, a stranger to approach a body in space near enough; it will inevitably raise tides in the other's mass, and if the approach be very close, the tides will be so great as to tear the body in pieces along the line due to their action; that is, parts of the body will be separated from the main mass in two antipodal directions. This is precisely what we see in the spiral nebula[e]. Nor is there any other action that we know of which would thus handle the body . . .
>
> As the stranger passed on, his effect would diminish until his attraction no longer overbalanced that of the body for its disrupted portions.

> These might then be controlled and forced to move in elliptic orbits about the mass of which they had originally made part. Thence would come into being a solar system, the knots in the nebula going to form the planets that were to be.[1]

Ironically, it would be Lowell's own assistant, Vesto Melvin Slipher, an Indiana University graduate hired by Lowell in 1901 to take charge of a new spectrograph, initially for the purpose of vindicating Lowell's contested ideas about Venus's rotation, who would prove the baselessness of Lowell's speculations. Instead of confirming Lowell's (and others') speculations about the spiral nebulae being solar systems in formation, Slipher made the unexpected discovery that they were (mostly) receding from us at high speeds. This in turn contributed to astronomers' eventual recognition that the spiral nebulae are something far more consequential even than solar systems in formation – they are galaxies, vast conurbations of stars in their own right, involved in the general expansion of the universe. By the time the expanding universe was being recognized, the close-encounter theory of the planets was also falling by the wayside. The wispy entrails of the Sun pulled away during a close encounter would simply have been too tenuous to stitch together into planets.

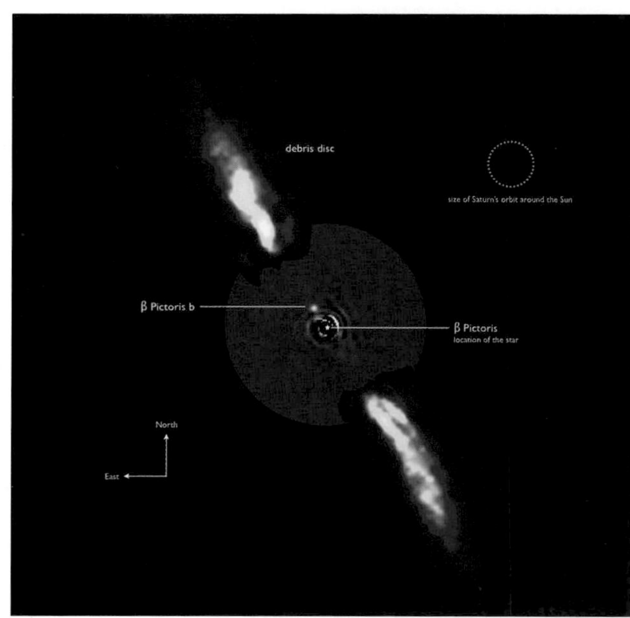

Kant and Laplace vindicated: false-colour image of a protoplanetary disc forming about the star Beta Pictoris, which is visible from latitudes south of Hawaii. The disc was discovered in 1984 by Bradford A. Smith (University of Arizona).

The Nebular Hypothesis Makes a Comeback

By the 1940s and '50s, the Nebular Hypothesis had returned to favour, as astronomers worked out ways of

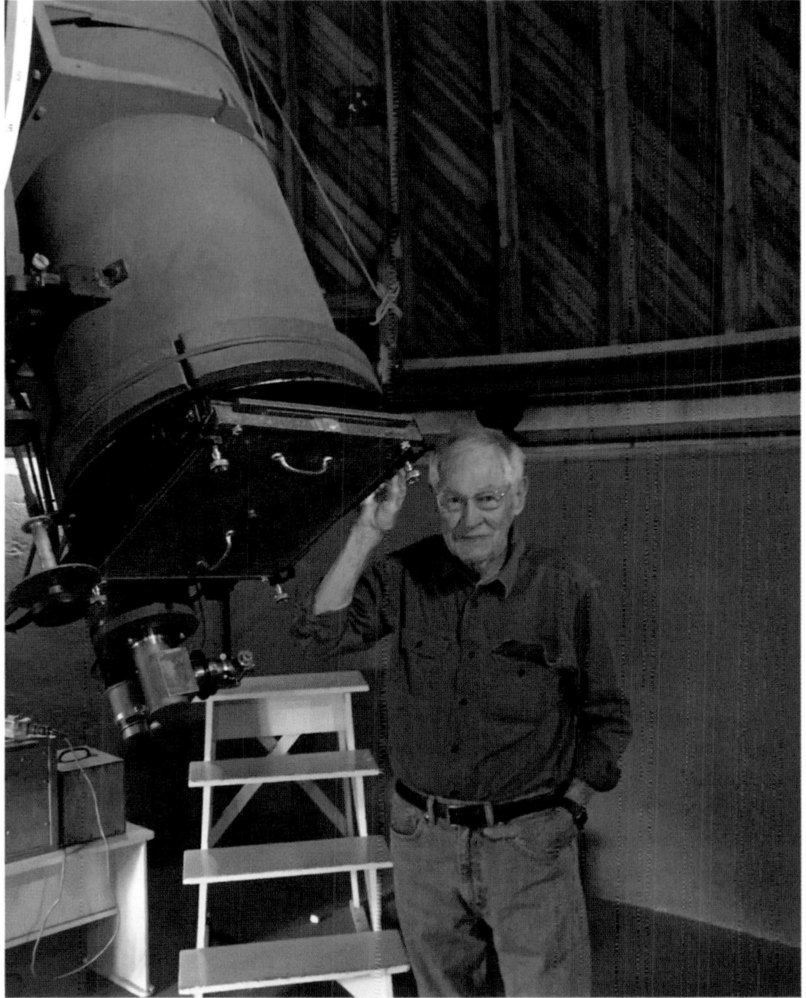

Bradford A. Smith, who early in his career was associated with Clyde Tombaugh's Planetary Patrol, a programme systematically carried out at New Mexico State University to photograph the planets. Smith went on to become Imaging Team Leader of the Voyager spacecraft missions to the outer planets, which included flybys of Jupiter in 1979 and 1980. After the Voyager 2 flyby of Neptune in 1989, he made the observations that showed the existence of the protoplanetary disc around the southern hemisphere star Beta Pictoris. Here he poses in 2016 with the 33-cm Abbott Lawrence Lowell astrograph used by his mentor Clyde Tombaugh to discover Pluto in 1930.

transferring angular momentum from the embryonic Sun to the planets. It is now universally agreed that the planets, satellites and other bodies of the solar system formed from a welter of debris left behind from the eddying protoplanetary disc of gas and dust surrounding the fledgling Sun, which must have looked very much like that now surrounding Beta Pictoris. It follows that planetary systems must be very commonplace throughout the universe –

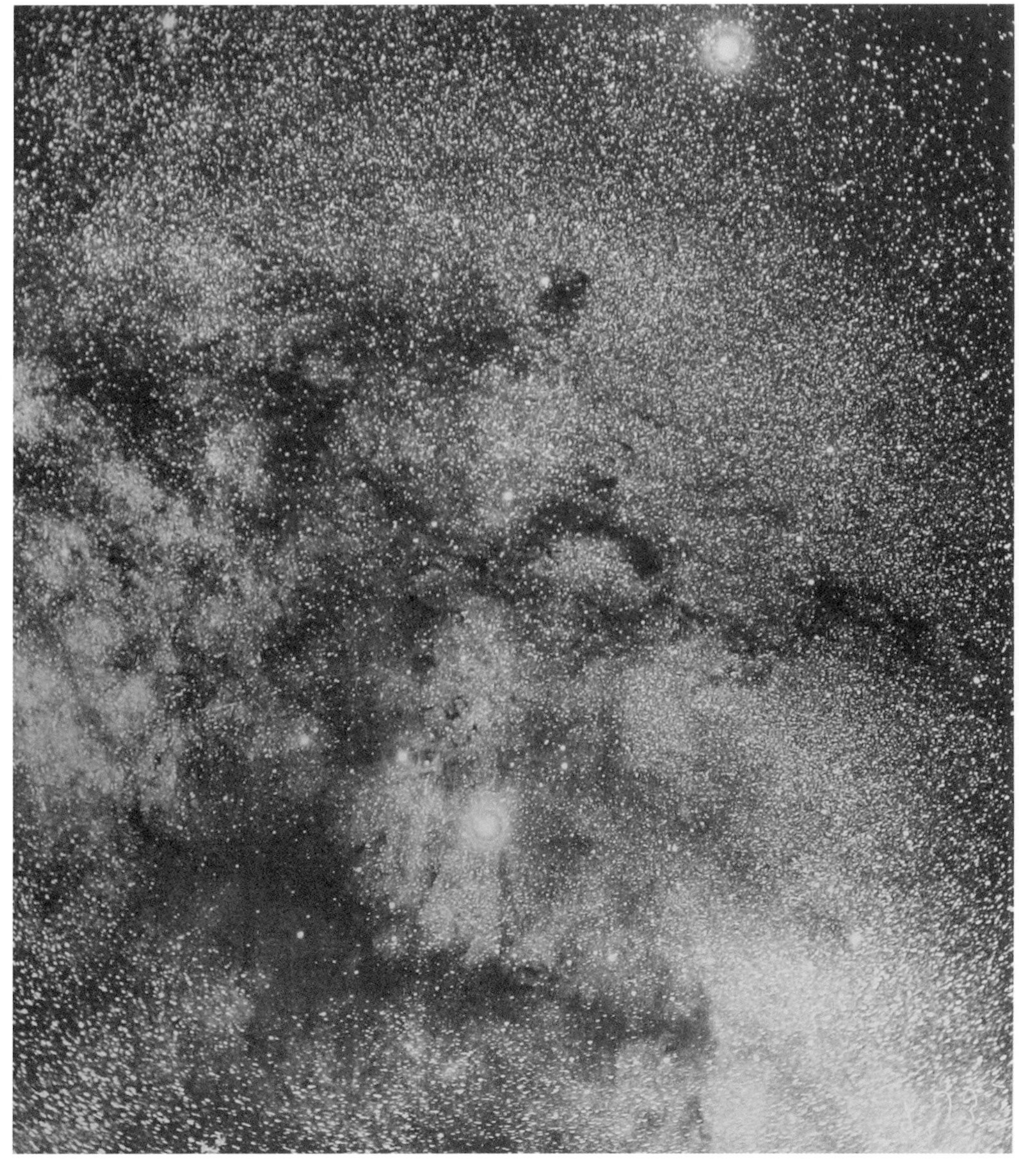

Dark clouds of the Milky Way. One of E. E. Barnard's classic Milky Way images, of 1913, taken with the 15.25-cm Willard lens of the Lick Observatory on 25 June 1892, exposure 4 hours. This image shows the region north of Theta Ophiuchi, with the 'Pipe Nebula', one of Barnard's dark nebulae, in the lower part of the image.

and so they are. Since the first exoplanet (a so-called hot Jupiter, around the star 51 Pegasi) was discovered in 1995, thousands of exoplanet systems – many strikingly different from our own solar system – have been discovered, and there is no end in sight.

A great deal is now known about the drama of the origin of the solar system, and after only the Sun itself, Jupiter has always played the leading role. In the beginning, 4.6 billion years ago, the Sun and planets emerged out of cold, dark, interstellar molecular clouds. Examples of these clouds were first recorded in the wide-angle photographs of the Milky Way taken by the great American astronomer Edward Emerson Barnard around the turn of the twentieth century. They were later studied by (and named for) the Dutch-born astronomer Bart Bok, who personally thought they should be called 'Barnard globules'. We see them as they are silhouetted against the background stars as we look from our position on one of the galaxy's spiral arms towards the centre of the Milky Way. These clouds are very cold, with typical temperatures of around 10 Kelvin (10° above absolute cold) and with densities of several thousands of molecules per cubic centimetre.

If a dark cloud is dense enough, or if it happens to be suddenly compressed by passage through the dusty arms of the Milky Way or by a supernova blast in its proximity, it begins to collapse in on itself. At first this collapsing tendency is resisted by the presence of magnetic fields, but eventually the magnetic fields 'leak out' of the cloud. Once this stage is reached, collapse begins in earnest, and the gravitational energy of collapse is converted into heat. However, because of the cloud's low temperature and low density, at first it remains transparent to radiation. The radiation simply escapes into cold space. The collapse at this stage is said to be isothermic – that is, it occurs without warming the cloud. During this isothermic collapse phase, the cloud undergoes fragmentation into hundreds of sub-clouds, each massive enough to contract further in its own right. These sub-stars, or protostars, continue to contract until

Looking along the plane of the Milky Way, the central bulge is visible (located in the direction of the constellation Sagittarius), and the plane is thick with the great dust clouds in which the stars and their systems of planets are formed, 2009.

they have become stellar-sized. Stars are what they are destined to become.

Though most of the material of the cloud falls directly into the incipient star and adds its mass, some leftover remnants are spun into a disc of the kind Kant and Laplace envisaged long ago. The swirling cloud of gas and dust out of whose broken rings planets form derives its symmetry from the same cause as the spiral form of the galaxy itself. It is the symmetry of matter in rotation. The gathering together of matter, realized on a grand scale in the starburst of the galaxy, is writ small in the formation of the solar system.

As the sub-clouds continue to contract, they rotate faster and faster, like a skater speeding up by pulling her extended arms inwards against her chest. In time the swirling material reaches supersonic speeds. The centrifugal force causes material in the cloud to flatten towards the outside, until the cloud's shape resembles that of a barred spiral galaxy. At this point it has formed a circumstellar disc, like that which has been imaged around Beta Pictoris. The rapid rotation leads to further fragmentation of the cloud, with most of the angular momentum of rotation becoming stored in the relative motion of the largest fragments. These fragments will go on to form the stellar components of binary- or multiple-star systems. Uncompanioned stars such as the Sun are much less common. (Incidentally, there is nothing to prevent binary stars from having their own planetary retinues and as we now know, most of them do.)

As the contracting cloud grows more and more dense, it finally reaches the point where it becomes opaque to radiation. No longer able to escape into space, the gravitational energy that is released as the cloud continues to collapse rapidly warms the womb-like interior of the cloud. As it does so, there is a corresponding build-up of gas pressure resisting further contraction. The opposing forces – gravity inwards, gas pressure outwards – eventually reach a delicately crafted compromise as the gravitational energy of collapse comes into precise balance with the heat energy of expansion. A wonder of nature – in basic structure elegant and simple but intricate in detail – appears. A star is born.

At first the star is a stellar pupa, tucked away inside its gas and dust cocoon, glowing with a softly beating irregular light. It is then known as a T Tauri star (after its prototypical namesake in the constellation Taurus; such a star has strong emission lines in the spectrum and a rapidly varying output of infrared, optical and ultraviolet radiation). Still surrounded by a gas and dust disc, such a star undergoes periodic outbursts, each lasting about a hundred years, in which mass is transferred from the disc onto the young star, increasing its luminosity. The gaseous component of the disc lasts only about 1 to 10 million years against the ravages of these periodic outbursts. However, it takes tens of millions of years – in the case of a one-solar-mass star, about 40 million – for the star to settle into an even-tempered luminary that will burn steadily and predictably for billions of years on the band of stars, graphed according to mass and brightness, that astronomers refer to as the Main Sequence. This means that by the time the star forsakes its turbulent youth and sets out as a stellar debutante on the cosmic stage, it has already begun to form a retinue of planets – or at least of gas-giant planets, such as Jupiter and Saturn. Calculations show that these giant planets must have formed within only the first 1 to 10 million years after the Sun itself, before the gas – hydrogen and helium, formed in the Big Bang and the chief components of the solar nebula – was dispersed into interstellar space.

At present, there are two competing theories regarding the formation of Jupiter and the other giant planets. The first assumes a top-down process, in which Jupiter formed through the very rapid and direct collapse of a cold, dense clump of gas and dust in the outer part of the circumstellar disc. Because the collapse occurs so rapidly, an enormous amount of heat would be trapped deep within the planet. The second theory assumes a bottom-up process, the so-called core-accretion model. Here, planet-sized cores of ice and rock form first and proceed to grow rapidly through an influx of gas and dust. A few objects – by a process that is fundamentally random and unpredictable, hence sometimes referred to as a Monte Carlo process – decisively outgrow the rest, and become the most massive and dominant objects in the solar system after only the Sun itself. This process is more gradual than that involved in a rapid and direct collapse, and the interior of the planet produced would be relatively cooler than in the top-down model, though still red-hot.

In principle, it should be possible to decide between these theories by observing infrared radiation from giant planets belonging to other stars, which are only a few hundreds of millions of years old. These planets have not yet had time to cool, as Jupiter and Saturn did long ago, and so should disclose the manner of their formation. Investigations along these lines are now in progress.[2] This is one avenue of approach. Another – much closer to home – is to observe Jupiter itself from orbiting spacecraft. As we shall see later, instruments on board the Juno spacecraft are currently gathering data to try to provide definitive answers regarding how Jupiter formed and evolved.

Whatever the details of its formation, Jupiter, once formed, would have been very difficult to destroy, and would not have been seriously affected even by the energetic outbursts that the young Sun experienced.[3] Youthful as it appears, with violent and tempestuous changes in its colourful clouds, it is in fact a very old world – at almost 4.6 billion years old, it is nearly as ancient as the Sun itself.

First Among Unequals

Other cores – embryonic planets – were forming at the same time as the one that formed Jupiter. The latter just happened to be situated in the right place at the right time to swell to unusually gargantuan proportions. The process of growing a giant planet from the bottom up is described in the title of a recent technical article, 'Growing Gas-giant Planets by the Gradual Accumulation of Pebbles'. That title conveys an astounding thought. The 'pebbles' were primordial centimetre- to metre-sized objects swirling in the protoplanetary disc, and out of these 'seeds', a gas-giant planet would form and grow to have a mass 318 times that of the Earth. The only thing that stopped it from growing further was the fact that it had succeeded in sweeping out all the gas along its orbit.[4] There was effectively no more substance left to add to its bulk.

According to a proverb, variously expressed, 'From small beginnings grow great things.' It was certainly so in the early solar system. What played out was akin to a giant game of Monopoly, in which the game board was on the scale of the solar system, and where the laissez-faire competition was not for money or land holdings but mass.

Meanwhile, in the inner solar system, which was largely depleted of gas early on, a variation of the same theme was playing out. There, bits and pieces of rocky debris formed the smallish lumps of the terrestrial planets, including Earth. From the outset the basic scheme of the solar system was thus established, each world being defined by the circumstances of its birth: small rocky planets inwards, giant gaseous planets outwards from the Sun. Our own position is close to the hearth. Compared to our gigantic brethren, we, with the other terrestrial planets, are little more than moths circling the flame.

Gravity's Grave Effects

All the planets revolve about the Sun in roughly the same plane, the ecliptic, so called because ancient observers noted that eclipses of the Sun and the Moon occurred exclusively in this plane. The direction of motion in their orbits is anti-clockwise as viewed from our north.

Though Jupiter now orbits the Sun in a nearly circular path at a mean distance of 5.2 AU (where 1 AU is an Astronomical Unit, defined to be the distance from the Earth to the Sun), it did not actually form in this position. As was realized by practitioners of celestial mechanics going back to Isaac Newton, the planets perturb one another, and over time the elements of their orbits change. It used to be supposed that these changes were bounded within reasonable limits; thus the great eighteenth-century practitioners of celestial mechanics, such as Joseph-Louis Lagrange and Pierre-Simon Laplace – the latter of Nebular Hypothesis fame – devoted a great deal of attention to demonstrating the 'stability of the solar system', in which the planets continued to follow more or less the same orbits forever, rather like a watch that would never run down.

It turns out, however, that things are rather more complicated than was once believed (and when is it ever otherwise?). As has

A scale model of the solar system from T.E.R. Phillips, ed., *Hutchinson's Splendour of the Heavens* (1925). This is a variant on a classic demonstration in introductory astronomy classes, in which the distances of the planets are represented to scale. Here, the planets are set in proximity to London landmarks: the Earth is over Boadicea's Statue, Mars over Scotland House, Jupiter over the turret of County Hall, Saturn over Adelphi Terrace, Uranus over St Clement Danes, and Neptune over St Paul's Cathedral. On the same scale as used here, the nearest star to the Sun, Alpha Centauri, would be in New Zealand.

become evident over just the last twenty or thirty years, the giant planets, Jupiter, Saturn, Uranus and Neptune, owing to mutual gravitational perturbations, have migrated around the solar system. As they have gravitationally shoved each other about they have wreaked havoc with the orbits of their planetary brethren. One result was to toss icy bodies from the inner solar system into the outer zones of the Kuiper Belt and Oort Cloud, the frigid regions from which periodic and long-period comets come. They have also had effects on the formation of terrestrial planets: Mars formed out of the part of the solar nebula which lay just inwards of Jupiter. Depleted of material like an embryo starved for nutrients, proto-Mars remained small, while between it and Jupiter the protoplanetary nebula that should have given rise to yet another planet remained unformed, and exists as the uncoalesced rocky debris that makes up the asteroid belt. The giant planets also threw a share of material into the inner solar system, producing the massive battering of the inner planets known as Heavy Bombardment (between about 4.1 and 3.8 billion years ago). The Moon, and Mercury and Mars, still bear the scars of this violent period in the form of heavily cratered terrain. (Although the Earth itself was no doubt subject to the same battering, because of weathering and plate tectonics only a few of the more recent impact formations survive.)

The first person to question the long-held assumption about the stability of the solar system – and the steadfast positions of the giant planets – was Renu Malhotra, now at the University of Arizona, who in the 1990s became interested in accounting for the rather bizarre orbit of Pluto, then still classified as an ordinary planet. Malhotra, born in New Delhi in 1961, was the daughter of an aircraft engineer at Indian Airlines. After completing a master's degree in physics at the Indian Institute of Technology Delhi in 1983, she went to the United States and entered the PhD programme at Cornell University. Her dissertation, completed in 1988, was on the moons of Uranus.

She obtained a post-doc at Caltech under Peter Goldreich, who with Scott Tremaine proposed in 1979 that Uranus's narrow rings were confined by a series of shepherd satellites. (Their theory was first confirmed in the case of Saturn's narrow F ring, which is bounded between two satellites, Prometheus and Pandora, unknown before the Voyager flybys; then, in the case of Uranus, by Voyager 2 during its flyby of that planet in 1986.)

When she began her study of Pluto, Malhotra noted that in contrast to the orbits of the other major planets, which are well separated, nearly circular and almost coplanar, Pluto's is highly irregular. Pluto's distance from the Sun at perihelion differs by almost 20 AU from its distance at aphelion, so for twenty years around perihelion (last passed in 1989) it lies inside the orbit of Neptune. It also makes excursions of 8 AU above and 13 AU below the plane of the ecliptic – one of the circumstances that had made it elude planet-searchers in the early twentieth century. Additionally, with its large satellite, Charon, it is effectively a binary planet. So the question was, how did this binary planet in its very peculiar orbit in the outer reaches of the planetary system form? And how, once formed, did it manage to remain in a stable position?

In the present epoch, because of a 3:2 orbital resonance with the outermost of the giant planets, Neptune, Pluto is protected from pernicious gravitational interactions. (The resonance involves Pluto completing exactly two orbits every time Neptune finishes three.) At one time it was thought that Pluto might be an escaped satellite of Neptune, ejected presumably in interactions with Neptune's large moon Triton (which moves in a backward or retrograde sense) into its present orbit. However, as knowledge of the characteristics of the Pluto-Charon system improved, the escaped satellite scenario came to seem decidedly implausible. Instead it seemed likely that Pluto had formed in an ordinary circular low-inclination orbit, along with many such small icy planets in the outer solar system.

At the time it formed, Pluto presumably lurked well outside the current 3:2 resonance with Neptune. But then havoc ensued. The giant planets – Jupiter, Saturn, Uranus and Neptune – which hitherto had moved in stable resonance positions were yanked out of those positions by gravitational disturbances produced by the belt of icy material in the outer solar system (now known as the Kuiper Belt) to which Pluto itself belonged. Saturn, Uranus and Neptune shifted outwards as Jupiter moved inwards. During this period of giant-planet migration Pluto and Charon became lodged in their unusual orbit.

Though adumbrated by a few earlier astronomers, the migration of the giant planets was first laid out by Malhotra, clearly and definitively, in a paper on the evolution of Pluto's orbit in 1993. At first her proposal was greeted with considerable scepticism, but it was greatly bolstered by the discovery of exoplanet systems featuring 'hot Jupiters', beginning with 51 Pegasi b, only two years later.[5] The hot Jupiters were clearly giant planets, and they must have formed well out from their parent stars before migrating inwards to their present positions. There was no longer any reason to doubt that similar processes were at work in our own solar system.[6]

Indeed, according to the present view of the case, the giant planets' gravitational perturbations first cleared out their zones of formation by scattering the remaining mass of planetesimals, the primitive small circulating bodies out of which the planets were made. Some fraction of this mass now resides in the Oort Cloud of comets; most, however, has been entirely lost from the solar system.

A planetesimal scattered outwards gains angular momentum, while one scattered inwards loses angular momentum at the expense of the planets. It is the effect of this scattering of planetesimals by the planets that causes their orbits to evolve. Most of the inward-scattered objects would have been thrown into the zones of influence of the inner Jovian planets (Uranus, Saturn and Jupiter), while of the outward-scattered objects, some would have been lifted all the way into the vast outer spherical cloud of comet nuclei known as the Oort Cloud and others would have fallen back towards the inner solar system to be re-accreted or re-scattered. Jupiter, owing to its large mass, would have been most effective in throwing objects out of the solar system, and in particular would have removed most of the scattered Neptune-zone planetesimals. As Neptune encountered this welter of scattered objects, it would have gained orbital energy and angular momentum – in effect, its orbit would have expanded.

By hurling these objects across the solar system, Jupiter served as the original source of Neptune's additional angular momentum and energy; however, because of its much larger mass, its own orbit would in consequence have shrunk by only a small amount. Summing up, as a result of these interactions, Neptune would have migrated outwards, Jupiter slightly inwards.

This scenario has profound implications for the dynamical history of the primordial small bodies in the outer solar system, especially Pluto. As Neptune migrated outwards, a series of gravitational disturbances called 'resonances' would have swept across the outer solar system. (Here a resonance refers to the case of two orbiting planets having periods that are small-integer ratios of one another, such as 2:1, 3:2 and so on. An analogous case is that in which a person on a swing is given a small push at the same point in each cycle; the push is in resonance with the natural frequency of the swing oscillation, and a large-amplitude excursion of the swing results.) If Pluto had formed initially with an orbital radius such that the 3:2 resonance was the first major Neptune resonance to sweep by, Pluto (and Charon, which presumably had been captured earlier by Pluto) would have been locked into this resonance with Neptune. The result is that, within its 3:2 resonance zone, Pluto and Charon are effectively hermetically sealed from planetary perturbations; they are safely locked within their protected orbit.

Note that, as usual, Jupiter, with its huge mass, was the key player in the game of musical chairs or do-si-do that played out in the early solar system. As we shall see, Jupiter continues to steer objects round the solar system and in many cases keeps them from bothering the Earth.

TWO

Jupiter: A Primer for a Giant Planet

Wherever it might have formed, and whatever its past sojourns might have been, at present Jupiter is fifth in order from the Sun among the planets, beyond the four terrestrial planets and the abortive planet that Jupiter itself prevented from forming, the asteroid belt. Late on a moonless night, at opposition or not, Jupiter shines as a subtly yellowish 'star' as bright as any in the sky. It is steady, calm, untwinkling, majestic – the undoubted royalty of the solar system.

Of course, the light by which Jupiter shines is only reflected sunlight. Imperial it may seem, but it is, like all the planets, a mere lackey and footman to the Sun. Sunlight that arrives on Jupiter is 27 times dimmer than sunlight arriving on the Earth. So how can it be that Jupiter, being so far away, and reflecting such dim light from its mirror, stands out so splendidly in our sky?

The mirror analogy suggests the answer. There are at least three ways of amplifying the light reflected from a mirror. One is to bring the mirror closer to the source, obviously, but that is an impossibility in the case of Jupiter. The others are to make the surface more polished and reflective or to increase the mirror's size.

Astronomers describe how reflective a surface is in terms of its albedo. Albedo is simply the ratio of the light reflected by a surface to the light received by that surface. Jupiter's albedo is not the highest of all the planetary surfaces, at 0.52, but it is close. By contrast, the albedos of airless and rough-surfaced Mercury and the Moon are 0.11 and 0.12, respectively. For the Earth, it is just 0.36. Only Venus, shrouded in brilliant clouds, surpasses Jupiter's reflectivity at 0.65.

Albedo certainly factors in, but the primary reason for Jupiter's apparent brightness is its size. It is eleven times the diameter of the Earth and almost one-hundredth the diameter of the Sun. The Maya were not so far-fetched in referring to it as the 'Sun of the night'. If it is a mirror, it is one that presents an enormous reflecting area.

Seen from the Earth, Jupiter's distance varies as it travels from the near side to the far side of its orbit relative to the Earth. As a result, its apparent diameter ranges from a mere 30 arc seconds (when it is in conjunction, on the other side of the Sun) to a maximum of almost 51" of arc. It is thus always larger than Mars. (Mars never shows an apparent diameter of more than 25 arc seconds.) Though the maximum apparent diameter of Venus exceeds that of Jupiter, reaching 60 arc seconds, this occurs when Venus is at inferior conjunction, between the Earth and the Sun. It then shows its night side to us. As a superior planet, Jupiter's disc is never a crescent or even a half; rather, it is always nearly full, with a maximum phase defect occurring when the planet is at quadrature – that is, when it makes an angle of 90° relative to the Earth–Sun line. Even then, Jupiter is still 99.1 per cent illuminated, so the phase can barely be appreciated. (A 99.1 per cent illuminated Moon looks, to ordinary human sight, like a Full Moon.)

These circumstances in combination make Jupiter one of the most magnificent and rewarding objects for the telescopic observer, since it can be observed to advantage during much of each year. This is in contrast to Mars, for instance, which shines brightly near opposition but, like certain flowering plants, is a biennial – it comes to opposition only every two years and two months. Jupiter is a perennial, and does so every 399 days on average. Mars, after a brief period of glory, fades as it recedes from the Earth and moves towards conjunction on the other side of the Sun, but Jupiter remains splendid for much of every year. It truly is the planetary observer's 'Old Faithful'. It never relinquishes its mighty sceptre; it remains enthroned, majestic, unruffled, almost like a Sun amid the other planets.

Incidentally, as seen from Jupiter, the Earth would be an inferior planet, always clinging to the Sun's apron strings. In a telescope, it would be seen to go through a cycle of phases, from new to crescent to half to gibbous to new, but it would be challenging to pick out from the solar glare, as its maximum elongation from the Sun is only about 11°. (Compare Mercury as seen from the Earth, which reaches as much as 28°.) So the Earth, our home, and the most important planet in the universe to us, would hardly be noticeable to a Jovian observer, did one exist. We are humbled in the presence of the giant planet.

Size and Composition

Knowing an object's apparent size and distance allows its actual diameter to be calculated. In the past, the apparent size of a planet was measured at the telescope by means of a micrometer: the separation of spider-threads is varied by turning a calibrated screw until the planet's disc fits snugly between them. Alternatively, in the case of Jupiter, attended as it is by four large satellites, its diameter can be determined by timing the duration of their passages across its disc. If one has a value for the apparent diameter, the actual diameter can be determined in a straightforward way if one knows the planet's distance. This is now known to a very high degree of accuracy. (Obviously, the distance varies depending on where Jupiter is in its orbit relative to the Earth, but the calculations needed, though complicated, can be easily handled by specialists in celestial mechanics.) On doing the maths, one finds that Jupiter is indeed truly enormous. Represented to scale on a flat page, eleven Earths can be fitted across it side by side, with a little room to spare. Since the planets are actually three-dimensional, of course, on adding the third dimension, one arrives at an even more impressive statistic: the circumference, the distance round the planet's wide girth, proves to be greater than that from the Earth to the Moon! This means that a Jovian Magellan, completing a circumnavigation of his planet, would have to have had eleven times the endurance of the terrestrial one.

Pedagogues from time immemorial have taught their young charges about the otherwise difficult-to-comprehend scale of the cosmos by using scale models. Following in their pedagogical footsteps, we here propound a scale model of the solar system, with the planets arranged according to their distances from the Sun and scaled to volume. Thus if the planet Mercury is represented as a pea, Venus and Earth are grapes, while Mars is either a small grape or a large pea. Then comes Jupiter – a slightly flattened cantaloupe! Saturn is somewhat smaller, perhaps the size of a grapefruit, while Uranus and Neptune, which are of roughly the same size, are a lemon and a lime.

From a dynamical point of view – the way in which one body's gravity influences the motion of another body – the quantity of stuff, or mass, making up Jupiter matters, not mere size. Jupiter's mass is 318 times that of the Earth, making it by far the most massive planet in the solar system. Its nearest rival, Saturn, is not even close, with only 95 times the mass of the Earth. Indeed, if all the other planets – giant planets, terrestrial planets and assorted debris such as asteroids and meteoroids – were thrown together on one side of a balance, Jupiter by itself would still topple the scale to the other side. It is by every measure the dominant planet in the solar system. In fact, the solar system might well be described, as a first approximation, as the Sun and Jupiter, the rest being mere parings and detritus.

In the 1980s, during a rough alignment of the Jovian planets (an alignment used to slingshot the two Voyager space probes through the outer solar system), there was a great deal said about the effect that their combined gravitational pull would have on the Earth. This was referred to as the 'Jupiter effect', according to which all the planets pulling along one direction greatly accentuate the magnitude of any one. Needless to say, it was sheer nonsense. Despite Jupiter's huge mass, since gravity falls off with the square of the distance, its force falls off rapidly. The differential gravitational attraction of the Moon, which is close to us despite not being very massive, causes much more disturbance in the terrestrial realm than does Jupiter, including most notably the tides in the oceans. The theory that the Jupiter effect might lead to worldwide cataclysms was puerile – but it sold books (many more, no doubt, than books written by respectable authorities on the planets).

Despite a volume so vast that 1,300 Earths could fit inside it, the material of which Jupiter is made is not very dense. It is far less dense than the Earth, which is made up of rocky and metallic materials, and has a mean density – that is, mass per volume – of 5.5 g/cm^3. Though Jupiter is believed to have a similar rocky core at its centre, the lion's share of its huge volume is gaseous. The planet largely consists of the lightweight elements hydrogen and helium. Its average density is only about 1.33 g/cm^3, which is just a little more than the density of liquid water (1 g/cm^3). What this means is that if it were placed in a supersized bathtub large enough to contain it, the giant planet would almost float. (Saturn, with a density of just 0.69 g/cm^3, actually would float in water; not quite like a cork, however, which has a density of only 0.24 g/cm^3.)

Stuff that Worlds are Made of

At one time, it seemed as though the composition of the planets and stars would never be known. However, this changed after the middle of the nineteenth century with the introduction of the spectroscope. The spectroscope, by means of a prism or diffraction grating, spreads the light from a planet or star into a band of different wavelengths. In the case of a planet's spectrum, dark lines and bands appear that can be compared to those in laboratory spectra. Just as each person has a characteristic set of fingerprints, so each element has a characteristic set of lines. This makes it possible to work out the chemical constitution of other worlds by analysing their light.

In the case of Jupiter and the other giant planets, an important step in understanding their composition was taken by V. M. Slipher of the Lowell Observatory, who in the 1910s and '20s recorded dark bands, known as absorption bands, in their spectra. At first, the nature of these bands was completely unknown. However, in the 1930s a German-born astronomer, Rupert Wildt, who left Germany after Hitler came to power and spent most of his subsequent career at Yale University in the United States, identified them as consisting of the hydrogen-rich compounds methane and ammonia. Only a few years before, stellar astronomers had begun to realize that the stars were made up mainly of hydrogen and helium. Wildt thought it likely that the composition of the giant planets was similar. In that case, Jupiter must be round 88 per cent hydrogen and 11 per cent helium. This, according to Wildt, was the recipe to make a Jupiter.

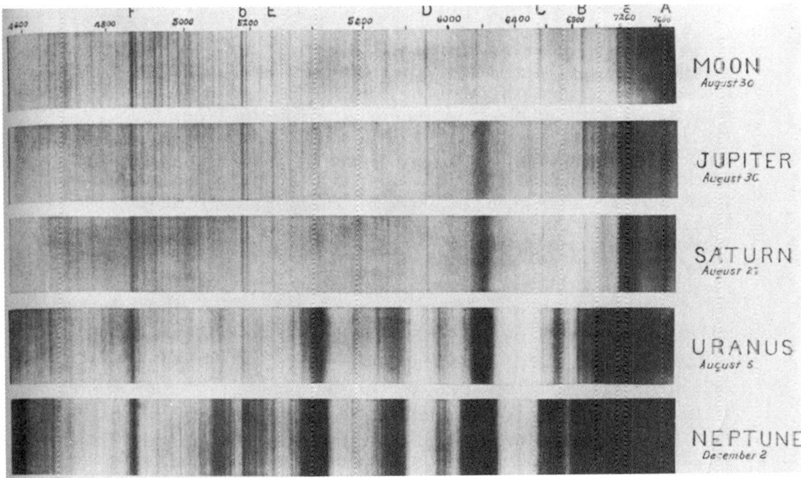

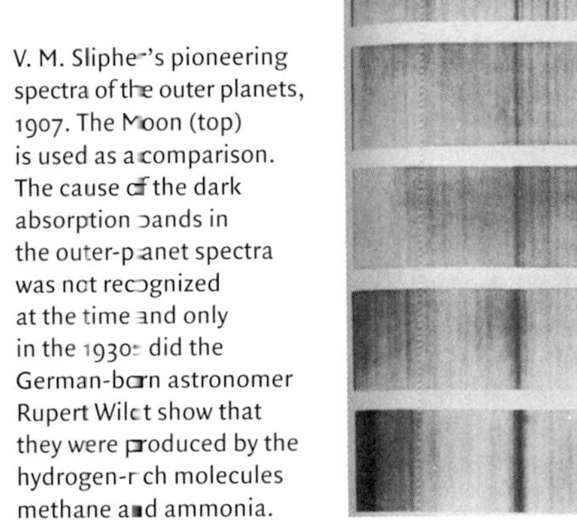
V. M. Slipher's pioneering spectra of the outer planets, 1907. The Moon (top) is used as a comparison. The cause of the dark absorption bands in the outer-planet spectra was not recognized at the time and only in the 1930s did the German-born astronomer Rupert Wildt show that they were produced by the hydrogen-rich molecules methane and ammonia.

Significantly, it also turns out to be the approximate recipe for the Sun. This is hardly surprising, given the way that Jupiter formed in the first place, growing rapidly by pulling in a large amount of the remaining gas in the solar nebula before it was dispersed. One might even say that the only real difference between a star like the Sun and a giant planet like Jupiter is the initial mass. If the mass is large enough, the gas will collapse into a star – that is, it will successfully initiate thermonuclear reactions in its core. If it is not quite large enough, it will become an almost-star, a giant planet. The next step, once the average composition and density of a planet are known, is to work out the internal structure based on the way that various materials behave under pressure. In general, materials deep below the surface ought to be highly compressed, that is, much denser than those closer to the surface.

In the 1940s and '50s, Wildt worked on this problem as well. His model of the internal structure of Jupiter – which, with some modifications, is still the basis of present-day models – assumed a large volume of hydrogen and helium gas surrounding a tiny embedded rocky core. Thus, in a way, Jupiter resembles a baseball, which consists of wool and cotton yarn tightly wound round a round cork centre. To complete the analogy, the visible cloud layers would resemble the stitched cowhide covering.

Jupiter: The Inside Story

Jupiter is not only the largest gas giant (and hence planet) in the solar system; it is also almost as large as it can be and still remain a planet. Add more mass to Jupiter and gravity will compress the resulting body into one with a *smaller* diameter. If the process were to continue, the pressure in Jupiter's core would soon become high enough to trigger thermonuclear reactions, where four hydrogen atoms are fused into one of helium, with the difference in mass released as energy according to the familiar formula $E = mc^2$. At this stage,

Jupiter would no longer be a planet; it would be a star (in which case the Sun–Jupiter system would represent a typical binary star).

Even so, Jupiter might be considered a kind of half-fledged or would-be star, since it releases almost twice as much energy as it receives from the Sun. Though a quantitative result for this excess energy output is based on quite recent measures using infrared detectors, as far back as the nineteenth century a number of astronomers, including Richard Anthony Proctor and Percival Lowell, intuited that Jupiter and the other giant planets might be embryonic stars. At least in the case of Jupiter, they were not far wrong.

Recall, again, the early history of the solar system, when Jupiter and the other planets were beginning to form out of the solar nebula. As matter began to stick together to form the first protoplanetary cores, and as some of these grew larger and attracted more matter, growing first to moon-like, then finally to planetary proportions, the collisions between them converted the energy of motion into the energy of heat. As Jupiter's rapid growth outpaced that of its rivals, its core grew hot. Much of this heat would be expected to remain bottled up – Jupiter is in fact a rather good, if outsized, Thermos bottle, with an exposed surface area that is small compared to its mass. On the basis of these considerations, Jupiter would be expected to retain at least some of its primordial heat. In addition to the heat produced during the early phases of its career as a planet, however, Jupiter may still be generating heat by ongoing contraction. Compressing a gas will heat it, as attested in the way a bicycle pump warms to the touch while being used. In order to retain its present energy output, Jupiter would have to shrink only a millimetre a year – hardly a rate that would be susceptible to detection by even the most diligent telescopic observer! Another possible source of heat may be helium rain – as it condenses and precipitates out of Jupiter's atmosphere, helium generates additional energy. (What a strange place Jupiter must be – a place where it rains liquid helium!)

An imagined view from Camille Flammarion's *Les Terres du ciel* (1884) of what the view on Jupiter might be like, looking sunward from a rocky surface through rafts of clouds. The dot on the Sun's disc is supposed to depict the Earth in transit. As we now know, there is no solid surface, as such, on Jupiter.

As a gaseous world – or gaseous-liquid world – Jupiter presents no solid surface to stand on. Instead, there are only clouds floating in the extended hydrogen/helium atmosphere, which become thicker as one plunges further into the planet's depths. At a depth of only a few thousand kilometres, the temperature and pressure increase to the point where hydrogen liquefies, something that at normal atmospheric pressures, as on the Earth, occurs only at extremely low temperatures; liquid hydrogen as we know it is a cryogenic material. On Jupiter, it is otherwise. There hot liquid hydrogen exists.

Though on a gaseous planet the term 'atmosphere' is somewhat ambiguous and imprecise, it will be used here to refer to a rather thin gaseous integument surrounding an underlying fluid ball. As has been known since the seventeenth century, Jupiter is both big and spinning rapidly. Indeed, its rotation, at just under ten hours, is the fastest of any of the planets. As a result, the fluid ball of Jupiter bulges at the equator, exactly in the way a bucket of water that is swung around will slosh to the bottom without losing a single drop. The force that holds the water against the bottom is referred to as 'centrifugal'– that is, centre-fleeing – force. Though we call Jupiter a gas giant, it might be more accurate to refer to it as a fluid giant.

The technical term for a ball flattened at the poles is 'oblate spheroid'. The amount of flattening is in general related to the rate of rotation. Thus in the case of extremely slow-rotating planets, like Mercury and Venus, the oblateness is zero. That of the Earth is 1/298, and of Mars 1/170. For Jupiter, the corresponding figure is 1/15. (Saturn, which rotates almost as fast as Jupiter, is even more flattened; its oblateness is 1/10.)

Because of Jupiter's essentially fluid nature, different layers within it rotate differentially. It is thought – though not yet proved – that the internal structure of Jupiter might be organized like a set of internested cylinders, with each cylinder rotating at a different speed. According to this view, the tops and bottoms of these cylinders make their presence known where they intersect with the 'surface' of the planet's globe and give rise to the different circulation regimes well known to observers of the visible cloud features on Jupiter.

Given that Jupiter's diameter is 143,000 km through the equator but only 134,000 km through the poles, the 'flattening' of Jupiter's disc is immediately evident using even small telescopes. A particle at the equator is whirled along at a rate of 45,000 km/h; by contrast, one at the poles has a rotational speed of zero. The differential rotation between the equator and the poles produces alternating circulation patterns, stretching the clouds into the gorgeously banded patterns that are a perennial delight to the telescopic observer.

Since Jupiter's atmosphere is thought to have been snatched from the solar nebula before it dispersed, it shares the composition of the solar nebula itself, and is primarily hydrogen, though it also contains small amounts of water vapour, methane and ammonia. Most of the original hydrogen and helium that the Earth possessed have escaped into space, but on Jupiter they have been held fast. There are two reasons for Jupiter's greater retentiveness. First, at the cloud tops of Jupiter, the temperature is –130°C, so molecules and atoms move sluggishly. Second, the planet's grip is remarkably tenacious. A measure of this tenacity is the escape velocity – the speed at which an object, whether a molecule or a rocket, must travel in order to break free of the planet's pull. Jupiter's is 60 km/s, compared to the 11 km/s of Earth. No rocket launched from the Earth so far has ever come close to achieving the escape velocity needed to escape the gravitational pull of Jupiter.

It is intriguing to imagine what it would be like were it possible to drop into Jupiter. As noted above, there is no solid surface to stand on, and anything that attempted to make a landing would embark on an endless – and fatal – descent. Of course, no astronaut will ever attempt it, but it can be done – and indeed has been done – by a robotic probe. On 7 December 1995 an atmospheric probe detached from the orbiting Galileo spacecraft and dropped into the clouds and down into the infernal Jovian depths. The exact point of descent was the south edge of the North Equatorial Belt. The exact spot of entry was 6½° north latitude, the location of one of the most ripping or high-speed jet streams on the planet. As the probe descended, it remained in contact for just under an hour; winds were found to increase at deeper levels, confirming that unlike the Earth, where cells of circulation are produced by solar heating, Jupiter's wind system is largely driven by internal heat, as described above. As the probe continued its soundings it found that, as expected, the pressure continued to increase at lower levels. Finally, at a depth of 155 km below the upper cloud deck, where the pressure increased to some 22 times that of the Earth at sea level and the temperature rose to 150°C, the doughty little probe finally succumbed to the extreme conditions. It had given its 'life' for science. At that point its radio transmitter went silent forever.

We can imagine the probe continuing its inward descent and ending up – who knows where. (Obviously, no matter what the manufacturer's warranty claimed, the metal and fibreglass making up the probe would have been utterly wrecked, reduced to vapour and mixed round with the rest of the swirling interior of Jupiter before it got much further down than the point where radio contact was lost.) Supposing, however, that it were made of sterner stuff and managed to reach all the way to Jupiter's core, it would have found itself among an undifferentiated amalgam of liquid rock, metal and water representing the primordial nucleus around which the giant planet formed in the early planet-forming sweepstakes 4.6 billion years ago. Here, at the very centre of the planet, the pressure is in the many hundreds of megabars (where 1 bar is the atmospheric pressure at the surface of the Earth), and the temperature is an estimated 30,000°C, which makes it much hotter than the visible 'surface' of the Sun. It would hardly be an inviting place – but then, for that matter, neither would the centre of the Earth.

THREE

SUPERFICIAL MATTERS

From what we have said so far, it is clear that Jupiter is a many-layered thing, and that the boundaries between gas and liquid are indefinite. We have said almost all we can about the vast and in many ways still mysterious interior of Jupiter, which contains by far the greater bulk of the planet's mass. Now we turn to the part of the Jovian structure that has been of absorbing interest to observers of the planet: the thin outer skin, which includes the visible layer of the cloud tops.

We are able to see only the upper 60 km of Jupiter's atmosphere. This is the realm to which the infinitely diverting and complex cavalcade of features that roll before the observer's eye on any given night belong. These features are the province of meteorology and can be understood in terms of its general themes.

A basic concept of meteorology holds that on a rotating planet – our own Earth, for instance – a parcel of gas warmed by the Sun will produce a column of rising air, which expands and cools. As it expands, gas on the equatorial side of the column must travel ever so slightly further to encircle the planet than gas on the polar side. As with horses in different lanes of a racetrack, the one with the longer distance to cover tends to fall behind. The effect of this lane differential is to produce rotation in the whole parcel of gas. Since the direction of rotation of the Earth is (as seen from above the North Pole) anti-clockwise, the rotation of a parcel of rising gas in the northern hemisphere will be clockwise (cyclonic), and in the southern hemisphere anti-clockwise (anti-cyclonic). (It may take a little doodling with a pencil and paper to convince oneself that this is the case. It might help to picture those horses running in different lanes.)

If the planet in question has a conventional axial tilt, again like the Earth, gas near the equator is heated more than gas near the poles. Thus the effect just described gives rise to the large-scale vertical cells of wind motion, which are well known to sailors. On the Earth, between latitudes 5° and 30°N and S blow the easterly trade winds, between 35° and 50° the temperate westerlies (the 'roaring forties') and at still higher latitudes the polar easterlies. At low latitudes are found bands of relative calm known as the doldrums (0–5°N and S) and the horse latitudes (30–35°N and S). (Note: winds are described by the direction from where not whither they blow; thus easterlies travel from east to west, westerlies from west to east.)

On Jupiter, which lacks a solid surface like the Earth, the reference point for atmospheric pressure is the upper cloud deck. The main source of energy is not the Sun but the heat supplied from within, and thus – strangely, from our point of view – it lies not in the sky above but in the depths below. Of course, sunlight still plays a role – there are small seasonal variations which follow the twelve-year Jovian year – but the convectively driven effects dominate and make those of solar insolation harder to see and measure.

Jupiter's internal energy source was strongly suspected, if as yet unproved, in the mid-nineteenth century. Despite its remoteness from the Sun – and the fact that the intensity of the solar radiation is only some 4 per cent of that on the Earth – many observers noted on its disc not the blandness expected of a world in deep freeze but the running motion of 'a windy sky', as full of clouds as the sky of the Earth on a warm summer day. In the words of the Scottish astronomer Charles Piazzi Smyth, who observed from the clear air of Tenerife, Jupiter showed the 'most picturesque clouds'.[1]

The formation of such picturesque clouds seemed to require an energy source supplemental to the enfeebled light of the Sun. So, in 1870, the English popular astronomy writer Richard A. Proctor wrote:

> That enormous atmospheric envelope is loaded with vaporous masses by some influence exerted from beneath its level. Those disturbances which take place so rapidly and so frequently are the evidences of the action of forces enormously exceeding those which the Sun can by any possibility exert upon so distant a globe . . . we seem led to the conclusion that Jupiter is still a glowing mass, fluid probably throughout, still bubbling and seething with the intensity of the primeval fires, sending up continuously enormous masses of clouds, to be gathered into

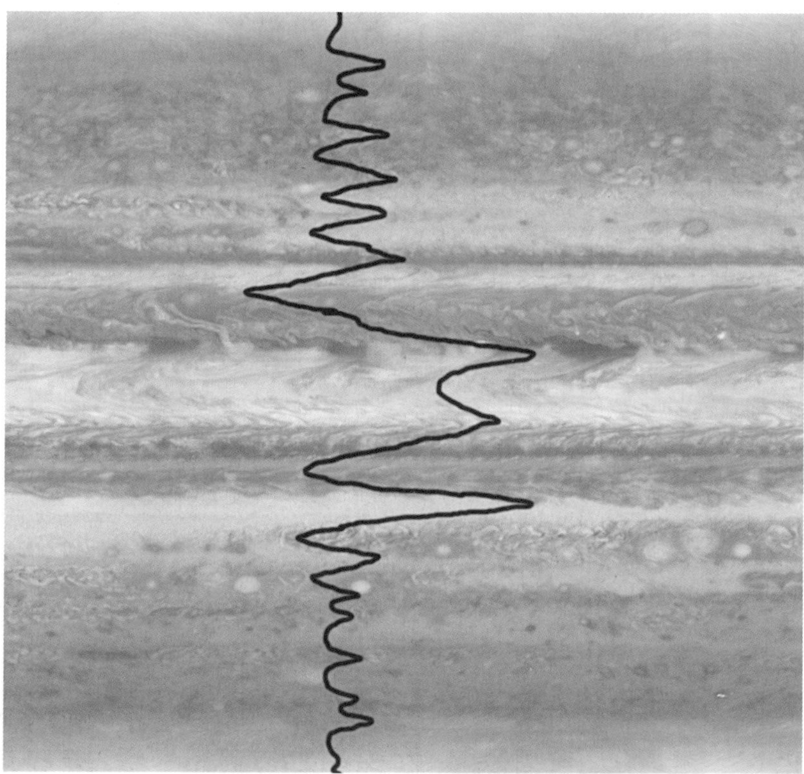

Jovian wind speed as a function of latitude, from NASA's Hubble Space Telescope.

'Windy Jupiter', as depicted by Piazzi Smyth, 1856. Appointed Astronomer Royal for Scotland, based at the Calton Hill Observatory in Edinburgh, in 1846, he is best remembered as a pioneer of astronomical observing at high elevations. With a grant from the Admiralty, he took a borrowed 10-cm Cooke refractor to Alta Vista, on the eastern side of Teide in Tenerife, which at 3,300 m was the highest point that mules could reach.

> bands under the influence of the swift rotation of the giant planet.[2]

Proctor's ideas were further developed in the early twentieth century by the American astronomer Percival Lowell, who wrote that the clouds of Jupiter:

> are not clouds ordered as ours. The Jovian clouds pay no sort of regard to the Sun. In orbital matters Jupiter obeys the ruler of the system; but he suffers no interference from him in his domestic affairs. His cloud-belts behave as if the Sun did not exist. Day and night cause no difference in them; nor does the Jovian year. They come when they will; last for months, years, decades; and disappear in like manner. They are *sui Jovis*, caused by vertical currents from the heated core and strung out in longitudinal procession by Jupiter's spin. They are self-raised, not sun-raised, condensations of what is vaporized below. Jove is indeed the cloud-compeller as his name implies.[3]

Percival was in this case prescient. It would be hard to find any part of this statement that has not been validated by subsequent research.

As hot gas rises in columns from below, it loses heat by condensation and produces eruptions of clouds, which the differential rotation swirls into vortices – cyclones and anti-cylones. In the polar regions, the columns are fairly well-preserved, and present as countless circular patches. In more moderate latitudes, some features, like the Great Red Spot and the White Ovals (see Chapter Six), remain true to their vortex nature over very long periods of time. Most are smaller in scale and more ephemeral, twisting into festively coloured festoons or stretching into thin wisps. The stretched features eventually contribute to the full-fledged bands and zones that extend all the way round the planet.

The overall scheme, then, is clear enough. But it is now time to leave generalities and introduce the specifics of detail that alone add life and colour to the subject. We turn to the actual observational history of the planet, and consider the remarkable record of Jovian atmospheric phenomena recorded by centuries of dedicated Earth-based observers, most of whom have been amateurs. The record indeed is incredibly rich. Because of the favourable circumstances that make Jupiter always such a large and attractive object for observation, no other planet has so much to offer.

FOUR

Atmospherics

Galileo Galilei was the first person to use a telescope to look at Jupiter, in the winter of 1610. At once he made a sensational discovery – the four largest satellites of the planet – a result that was more stimulating to his imagination than any other, since it seemed to show that Jupiter was the centre of its own system of planets, a little solar system arranged round a different centre from the Sun. However, his telescopes, though the best of their time, were not perfect enough to show any details on the disc.

It took another generation before the planet's characteristic banded pattern was made out by several Italian astronomers, including Niccolò Zucchi, Evangelista Torricelli and Francesco Fontana. Though these observations, made around 1630, marked the limits of what the telescopes of their day were capable of – the furthest verge of science that early seventeenth-century optics could reach – they can be easily emulated with just about any reasonably good small modern telescope, even most of those of the notorious 'department store' variety.

The observational record compiled over the nearly four centuries since Fontana produced his exquisite little 'Giove' shows that the banded backdrop is, though not entirely, at least reasonably stable. Thus some of the Italo-French astronomer Giovanni Cassini's and the Dutch astronomer Christiaan Huygens's drawings in the 1660s and '70s are virtually indistinguishable from those made with small telescopes today.

Were the Jovian vistas entirely stable, however, they would soon cease to be of interest. It is the possibility of change, of movement and novelty, that makes any study attractive, and beneath Jupiter's steady-going facade is a restless and dynamic panorama, with the smaller features constantly undergoing change and even the broader bands and zones subject to dramatic differences in hue,

The first page of Galileo's journal of observations of Jupiter's satellites, based on two earlier sheets for the night of 15 January 1610.

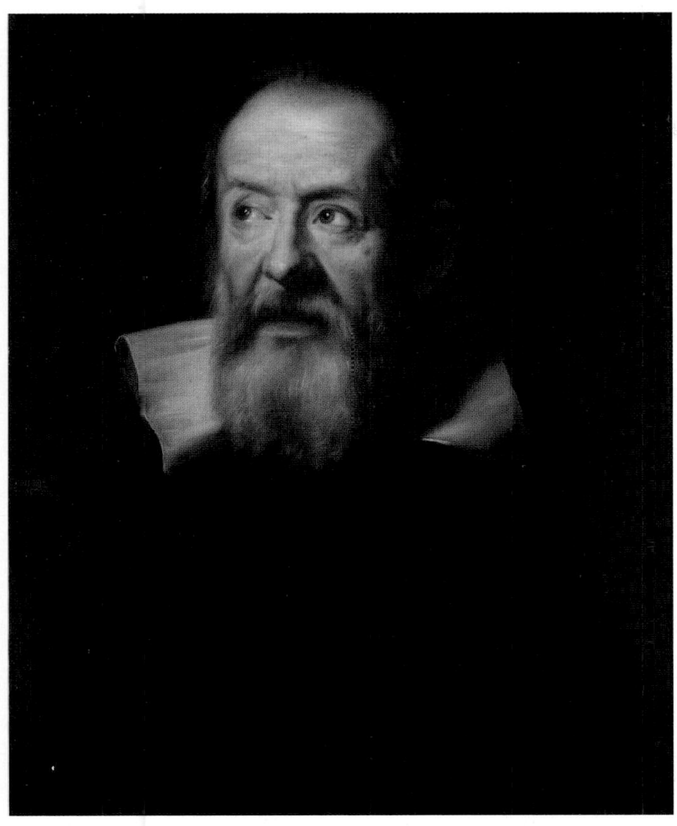

Justus Sustermans, *Galileo Galilei*, 1636, oil on canvas.

brightness and appearance. Also, because of the frantic pace of Jupiter's rotation, there is plenty of variety on a single night. Except during the abbreviated nights of midsummer, an observer, with stamina duly fortified at intervals by helpings from a pot of coffee, may peruse, in dizzying array between dusk and dawn, the entire globe of Jupiter! Nor does Jupiter ever look exactly the same night to night, since we are not looking at solid ground, as on the Moon or Mars, but at the mutable and changing cloud features of an atmosphere. Truly, old Jupiter, staid and majestic as it may seem, is in its cloud deck a world of evanescence and mutability, and every night brings something new and fresh.

It is hardly necessary to emphasize again that in telescopes big or small, our view of Jupiter is limited to the canopy of multicoloured cloud layers floating in the uppermost strata of its atmosphere. These colours, which make the planet so gorgeous in the telescope, are themselves instructive as to the nature of those clouds; we do not need a spectroscope to make deductions about their chemical composition but can do so through eyeball colourimetry. Compounds known as chromophores that are present in trace amounts produce the colours. Hydrogen and helium, which are Jupiter's main constituents, are completely transparent gases;

Christiaan Huygens (1629–1695).

helium, moreover, as a noble gas, is unsociable or – as chemists say – inert. It does not combine with other atoms to form molecules. Hydrogen, on the other hand, has the opposite temperament. It is exceedingly sociable and gregarious, and readily combines with any other elements that happen to be around, such as oxygen, carbon, nitrogen or sulphur, which make up less than 1 per cent of the planet's bulk. In combination with hydrogen, these elements form molecules such as water, methane, ammonia, hydrogen sulphide, ammonium hydrosulphide and various organic compounds. (Note that 'organic' here does not imply that they are produced in living organisms, only that they involve bonds with the versatile atom carbon.) It is these molecules that provide the ravishing hues that we see on Jupiter.

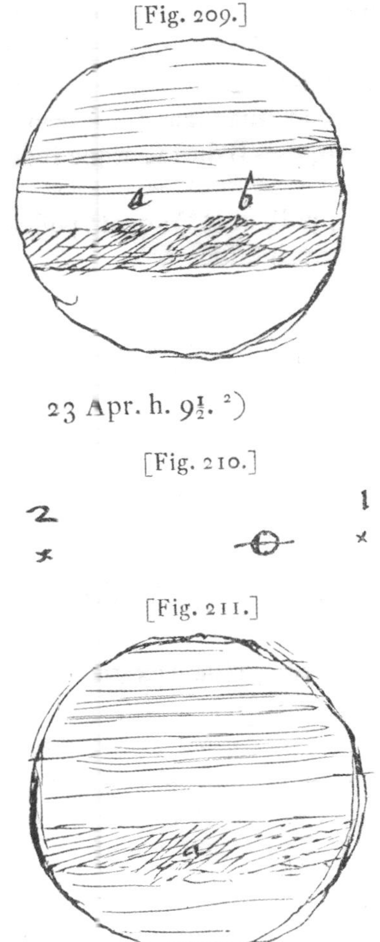

Sketches of Jupiter by Christiaan Huygens, 1685. These sketches very closely resemble those W. S. made with a small reflecting telescope in the 1960s. The small sketch, which looks like Saturn, shows Jupiter's globe and the plane in which the satellites move.

Apart from methane, which is always gaseous, a characteristic of these other molecules is that they condense under the unique conditions of temperature and pressure in Jupiter's atmosphere. At this condensation level, finding itself unable to rise higher, the condensate spreads out and forms a cloud deck. Thus the colours on Jupiter betray information about not only the presence of trace molecules but the temperature and pressure of the clouds in which they form.

At the level of the upper cloud tops, where the weather and chemistry are determined both by the energy of sunlight and by the much-attenuated heat upwelling from the interior, the temperatures are very frigid, about −130°C, and the clouds consist of ammonia-ice cirrus. Under equilibrium conditions ammonia cirrus would appear perfectly white; however, photochemical reactions involving ammonia and methane generate a yellowish-brown smog that accounts for the planet's characteristic yellow and tawny hues. The brownish belts are found at a somewhat deeper, warmer and higher-pressure level. Though it has not yet been definitively proved, it is likely that the ammonia ice above them has burned off so as to unmask the brownish ammonium hydrosulphide smog below.

It is generally agreed that the whitish zones are high-pressure areas where ammonia rises up and freezes into creamy ammonia cirrus clouds; the belts are low-pressure areas where downwelling prevents ammonia from rising and keeps the region clear, allowing a peek at the lower, warmer brownish clouds. Though on Earth we are also familiar with cells of high or low pressure, and approaching weather systems can be anticipated based on changes in barometer readings, on Jupiter – because of its rapid rotation – the high- and low-pressure regions spread into circumferential bands. Roughly a dozen bands – consisting of alternating light zones and dark belts – are visible at any given time. The flow in the belts and zones is in general far from smoothly laminar. Instead it is subject to a high degree of shearing and vorticity. Finally, on a planet where winds howl at half the speed of sound, it should surprise no one that the general appearance, in a telescope of sufficient aperture, is one of seething and roiling chaos. In beholding this planet, one might imagine oneself, as John Milton puts it in *Paradise Lost* (Book II):

> . . . the sport and prey
> Of racking whirlwinds . . . for ever sunk
> Under yon boyling Ocean, wrapt in Chains;
> There to converse with everlasting groans.

In more prosaic summary, we note that in the upper 60 km of the Jovian cloud decks we are able to see there exists a thin 'weather layer' in which the energy of the Sun, which drives the weather patterns on Earth, is pertinent only at the very top, and the rest of the circulation is driven by the heat upwelling from the deeper, warmer, fluid regions of the planet's interior. Though generally speaking the background is stable, consisting of bright zones and darkish bands, the detail is mind-bogglingly complex: feature is packed within feature in the manner of Chinese boxes. For a century and a half or more it has been the business of backyard observers (notably, the stalwarts of the British Astronomical Association, and more recently of the American Association of Lunar and Planetary Observers) to maintain careful and complete records of the weather of this other world from their private observatories and Jovian meteorological stations.

Some Notes on a Windy Planet

The darker belts and brighter zones are referred to by means of a convenient if rather perfunctory system of nomenclature first adopted by the observers of the British Astronomical Association (BAA, founded in 1890), and now too well established ever to be given up. The basic scheme is shown here. The usually bright (though sometimes ochre or even tawny) zone spanning 9° on either side of the equator is known as the Equatorial Zone (EZ), and the darker belts on either side of it are the North and South Equatorial Belts (NEB and SEB). (The SEB is frequently double – it presents as a pair of belts instead of a single belt – in which case the components are identified as SEB (N) and SEB (S)). At still higher latitudes are the North and South Tropical Zones (NTrZ and STrZ), the North and South Temperate Belts (NTB and STB) and so on. The scheme begins to break down beyond 45° North and South, where the banded pattern gives way to mottling due to convection cells.

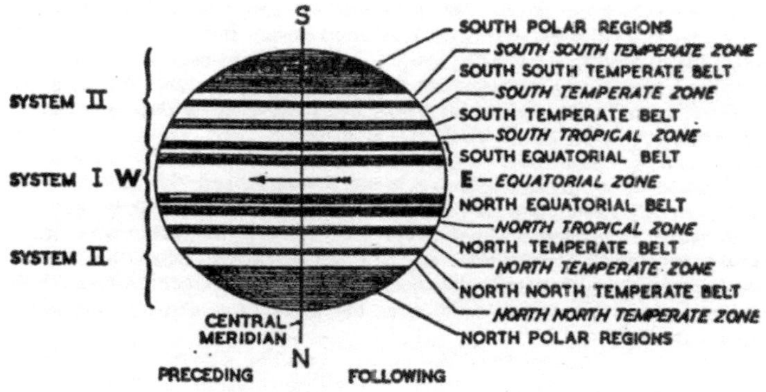

Nomenclature for Jovian belts and zones, first adopted by the British Astronomical Association.

The Paris Observatory, founded by Louis XIV, as it appeared during Cassini's time, from Camille Flammarion, *Les Terres du ciel* (1884). The tower on the right is the 'Marly Tower', a dismantled part of the Machine de Marly originally designed as a hydraulic system in Yvelines, France, built in 1684 to pump water from the Seine and deliver it to the Palace of Versailles. This version of it was commandeered by Cassini for mounting his long focal length 'aerial telescopes'.

Though, as mentioned above, the impression of bands was settled by the Italian observations of the early 1630s, it took another generation of improvements to the telescope before such notable observers as Robert Hooke, Christiaan Huygens and Giovanni Cassini began to detect irregularities in the bands and discrete spots distinctive enough to be recognized from one observing session to the next, and sometimes even from one apparition to the next. These irregularities and spots would prove to be useful in determining the planet's period of rotation.

Thus in 1664 Robert Hooke recorded a smallish spot, which used to be put forward as the earliest record of the famous Great Red Spot (though the latter did not receive its name until the nineteenth century). However, Hooke's spot is not in the right location; it lies on the NEB instead of on the edge of the SEB. A somewhat more convincing candidate is a spot seen in 1665 by Giovanni Cassini. He was still in Bologna, Italy, at the time but soon (in 1669) was enticed to France by Louis XIV or his chief minister Jean-Baptiste Colbert to take charge of the Paris Observatory. Cassini's spot is at least in more or less the right position, and seems to have been rather long-lived; indeed, its persistence led to its being referred to as the 'Permanent Spot'. Cassini himself recorded it on various occasions between 1665–6 and the early 1690s, and a feature, evidently the same and depicted as deeply red in colour, appears in a 1711 painting by the Italian artist Donato Creti now in the Vatican Museum. Creti spent most of his career in Bologna, and though Cassini himself never returned to Italy after 1669 – he died, completely blind, in Paris in 1712 – he had employed the artists Jean Patigny and Sébastien Leclerc in the preparation of his great Moon map. Creti's painting also gives the impression of having been based on direct telescopic observations. (Parenthetically, the 'Permanent Spot' painting was one of a series of small canvases depicting celestial bodies commissioned by the Bolognese count Marsili as gifts for Pope Clement XI. They were intended to motivate the papal states to found a public observatory, an effort that was ultimately successful and led to the construction of the tower observatory of the University of Bologna in 1725.)

As noted, Creti's painting dates from 1711. There is one more observation of the 'Permanent Spot', made by Cassini's nephew, Giacomo Maraldi, at the Paris Observatory in 1713, whereupon it disappears from the record. It is still sometimes identified with the

Donato Creti, *Jupiter*, 1711, oil on canvas.

Great Red Spot but its relatively small size – and its disappearance after 1713 – make this rather unlikely. It may well have been an earlier feature of the same kind.

Already in the seventeenth century astronomers debated the nature of the markings that presented to their telescopes. In his *Cosmotheoros*, published posthumously in 1698, Huygens discerned that Jupiter was a windy planet, and went so far as to discuss the consequences of its weather for the Jovian inhabitants:

> In Jupiter have been observ'd Clouds, big no doubt with Vapors and Water, which hath been proved by many other Arguments, not to be wanting in that Planet. They have then their Rain, for otherwise how could all the Vapors drawn up by the heat of the Sun be disposed of? and their Winds, for they are caused only by Vapors dissolved by heat, and it's plain that they blow in Jupiter by the continual motion and variety of the Clouds about him. To protect themselves from these, and that they may pass their Nights in quiet and safety, they must build themselves Tents or Huts, or live in holes [underground]. For I dare not affront the Pride of Men so much as to say, they are as good Architects, have as noble Houses, and as stately Palaces as our selves. And good now who are we? Why a company of mean fellows living in a little corner of the World, upon a Ball ten thousand times less than Jupiter or Saturn . . .[1]

Clearly, the clouds were in a state of continual motion and rapid change. In addition to various dark spots, the early observers registered bright patches from time to time and wondered whether they were not being treated to glimpses of the solid surface through gaps in the clouds. (But which was the ground and what was cloud?) Cassini, for his part, spoke of having seen the 'snow-covered hills' of Jupiter. A century later, even William Herschel, for a time, and Johann Schröter made the same mistake in assuming that the white

Johann Hieronymus Schröter (or Schroeter, as his name was always spelled on the title pages of his published works), the great German pioneer of lunar and planetary observations.

areas were the actual surface and the dark areas clouds. Not until the early nineteenth century was this figure/ground problem, to introduce a bit of psychological jargon, properly sorted out. As the reader who has remained with us thus far already knows, the bright, billowy clouds are located at a higher elevation in the Jovian atmosphere than the darker features. The early observers were wrong about this – and evidently also in their belief in Jovian inhabitants.

FIVE

The Great Red Spot Becomes Great

Though it has often been assumed that Cassini's 'Permanent Spot' of the late seventeenth century is identical with the now-famous Great Red Spot, this is far from certain. Giacomo Maraldi saw it in 1713, after which there is no further record of it for more than a century – a hiatus that would seem, in the words of British Astronomical Association Jupiter Section Director John Rogers, to represent 'an unbridgeable gap'.[1] It has sometimes been suggested that the disappearance of the 'Permanent Spot' may somehow be connected with the virtual lack of sunspots observed during the same period – the so-called Maunder Minimum. It is possible that the two are related, though we hasten to add that it hardly solves one mystery to conceal it in another.

For whatever reason, the records of even such diligent observers as William Herschel and Johann Schröter fail to give any indication of the Great Red Spot during the period covered by their observations (the late eighteenth century), though they do attest to many other interesting Jovian phenomena, and it is hard to see how they could possibly have missed it had it been there. The diligent Schröter, for instance, in 1785–6 observed an outbreak of what he referred to as 'schwarzdunkler Flecken' (extremely dark small spots).[2] These spots are likely to have been an early example of the so-called 'blue features' phenomena, that is, bluish projections emerging from the edges of the dark belts. Since 1911, they have always occurred on the south edge of the North Equatorial Belt (NEB), but before that, going back at least as far as Schröter's time, they appeared on the other side of the equator, on the north edge of the South Equatorial Belt (SEB). However, Schröter's rather extensive records give no indications of the Great Red Spot.

Amateur astronomer G. D. Hirst, in Sydney, New South Wales, Australia, produced this rather bizarre drawing of Jupiter, showing a feature he referred to as the 'Fish', on 9 May 1876. It appears this may have been an early stage of the emergence of the feature that came to worldwide prominence two years later as the Great Red Spot.

We can definitely trace what might be called the 'modern' history of the Great Red Spot only to 1831, when Heinrich Schwabe, a pharmacist in Dessau – today best remembered as the discoverer of the eleven-year sunspot cycle – recorded the telltale notch or embayment in the SEB known as the 'Red Spot Hollow'. Since Schwabe made this observation, the Hollow has always marked the Great Red Spot's position, even when the latter itself has been invisible. Thus the Red Spot (often faint) or Hollow are represented in subsequent drawings by Schwabe himself, the Revd William Rutter Dawes, William Huggins and Laurence Parsons (the 4th Earl of Rosse), made at intervals over the next forty years.

Up to this point, the Great Red Spot had not yet achieved any particular notoriety – or even received a specific name. However in 1876 – with the best observations of the planet being sent from New Zealand and New South Wales because of the planet's location far south of the celestial equator – the Sydney merchant and yachtsman G. D. Hirst, using a 27-cm silvered Newtonian, obtained drawings which show a rather bizarre bright reddish feature emerging from what is now referred to as the SEB. (Previously, Hirst had had difficulty seeing planetary colours, so his description is noteworthy.[3])

John H. Rogers, long-time director of the Jupiter Section of the British Astronomical Association and author of the definitive work on Jupiter of its era, *The Giant Planet Jupiter* (1995), 2015.

From the shape of the feature, Hirst referred to it as the 'Fish'. His drawing finally arrived at the Royal Astronomical

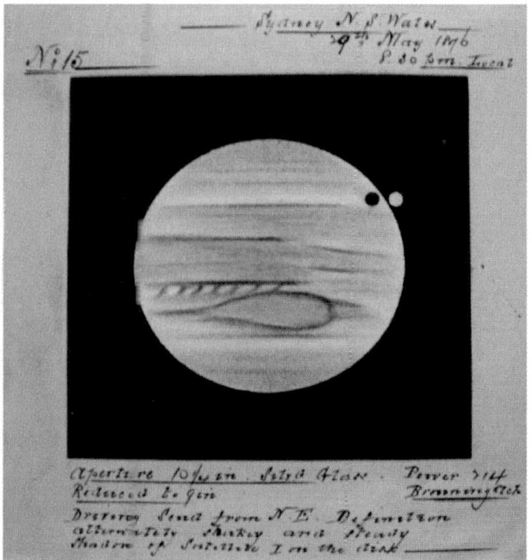

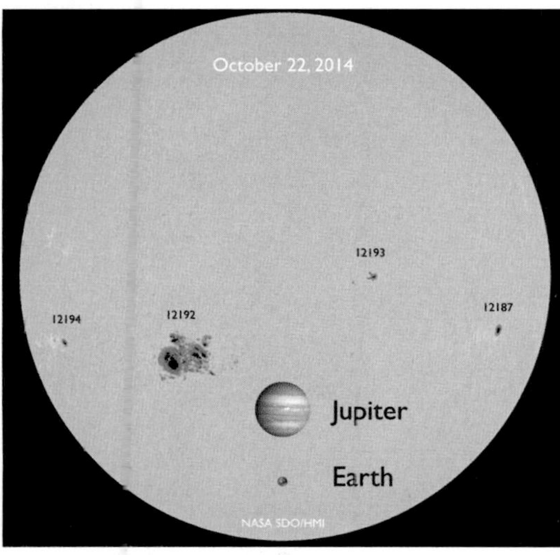

Sunspots present on the Sun, 22 October 2014, with the Earth and Jupiter to scale. The sunspot group designated 12192 was one of the largest of the cycle.

Society in 1877, and so impressed noted instrument-maker John Browning that the latter went so far as to frame it and exhibit it at the next RAS meeting. Little happened in 1877, and only in the following year did the 'Fish' – if in fact, as seems likely, it was the same feature – complete its rise to fame, and become, as the 'Great Red Spot', far and away the most celebrated feature on the planet, and a name which Thomas Babington Macauley's 'every schoolboy' would know.

On 9 July Carr Waller Pritchett, an ordained Methodist minister, first president of the now-defunct Pritchett College in Glasgow, Missouri, and director of the college's Morrison Observatory (founded with a donation from Berenice Morrison, whose maternal grandparents owned a tobacco plantation in Glasgow), was observing the planet with the 'world-class' 30-cm Clark refractor when he noted an 'elliptic cloud-like mass, separate from the general contour of the belts. This cloud was almost a perfect oval in shape,' he wrote, 'and was preeminently rose-tinted.'[4]

Pritchett's report generated tremendous excitement. From a mere handful of devotees, Jupiter now lured hundreds to telescopes to see what was advertised, in carnival barker fashion, as the 'greatest Jovian feature of all'. As it proceeded to darken to intense brick red, it began to be called by the name by which it has been known –

though only occasionally deservingly – ever since: the Great Red Spot. By 1880 its majestic oval outline, which for a time roughly approximated to the ellipsoidal shape of a cigar, reached 40,000 km × 13,000 km. To put this in perspective, three Earths could be fitted inside it side by side with room to spare.

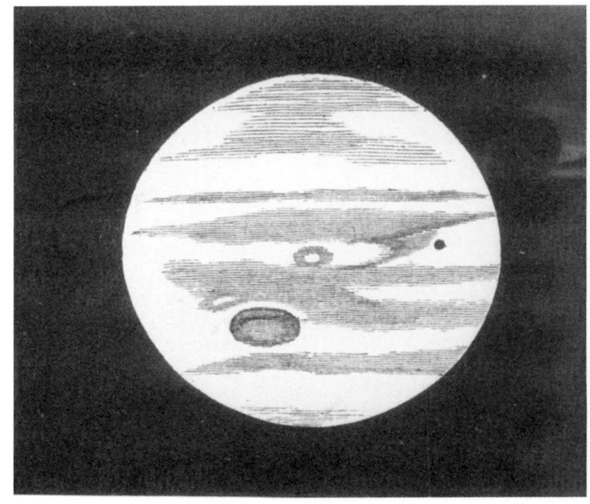

Jupiter, drawn on 29 July 1878 by Carr Waller Pritchett using the 30-cm refractor of the Morrison Observatory in Glasgow, Missouri (later moved to Fayette, Missouri, where it remains to this day).

Among the observers who were captivated by the Great Red Spot was Edward Emerson Barnard, then a young, self-educated photographer's assistant in Nashville, Tennessee, who possessed an unquenchable thirst for astronomical knowledge. Barnard made a careful series of observations with a 10-cm refractor acquired by prudent scrimping and saving – the telescope cost $380, which was a lot of money, a full two-thirds of his annual salary at the photography gallery at the time. Barnard was then living with his invalid mother in a 'house with a mansard roof', in a rather rough and desolate

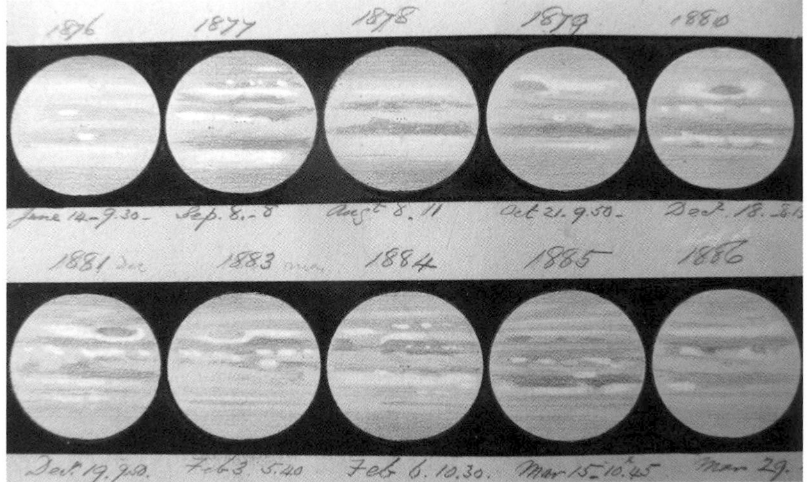

Using pre-prepared discs of creamy colour and with the appropriate oblateness, the artist-astronomer Nathaniel Green produced these drawings showing changes in Jupiter's cloud tops between 1876 and 1886. Note the elongated form of the Great Red Spot in 1879, 1880 and 1881, and its faded appearance in 1883–4.

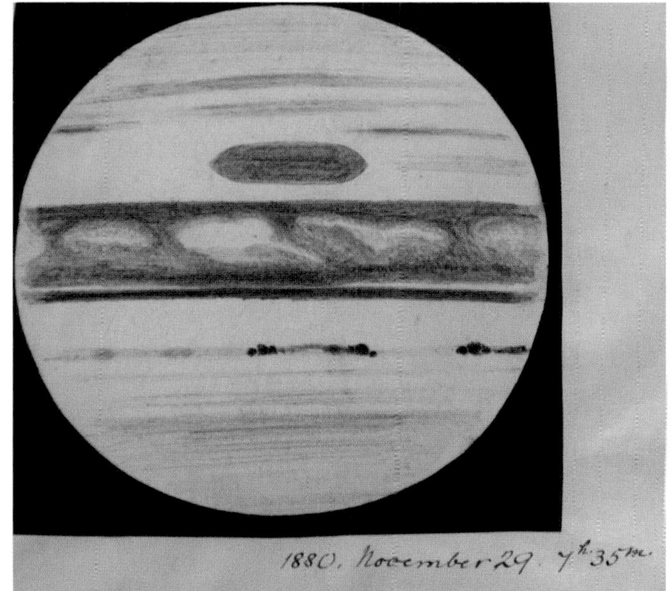

The Great Red Spot at its largest. Drawing by W. F. Denning, 29 November 1880. South is at the top, corresponding to the usual view in an astronomical telescope. Note the unusual series of minute dark spots in the North Temperate Belt.

Red Spot by astronomer-artist Nathaniel Green, 2 November 1881, made with a 33-cm reflector at St John's Wood, London. Green, an artist by profession, had tinted cards printed for his use in drawing the planets (ochre for Mars, cream for Jupiter). He could draw on these discs in soft pencil or pastel, then scrape away the creamy tint to give the highlights. At this time the Great Red Spot was nearly at its maximum elongation and redness.

A 23-year-old Edward Emerson Barnard sits for a pencil portrait with his pride and joy, a 10-cm refractor acquired with his savings as an assistant in a photograph gallery in Nashville. With this telescope, he observed Jupiter in 1879–80, and later achieved fame as a discoverer of comets.

patch of Nashville 'where there was only one neighbor near', and he afterwards recalled the loneliness of the place where he made his observations, which 'oftentimes impressed me with a kind of dread, for I was out at all hours of the night'.[5]

Barnard observed Jupiter throughout 1879 and 1880. In August 1879 he found the Great Red Spot the colour of 'red-hot iron'.

Meanwhile, it was undergoing changes in its oval form: the ends occasionally appeared rounded and at other times pointed or tapered into thin trails. The Great Red Spot was also found to repel other features that approached it.

After reaching its maximum length in 1880 and its deepest red a year later, the Great Red Spot began to fade in 1882, and for the next

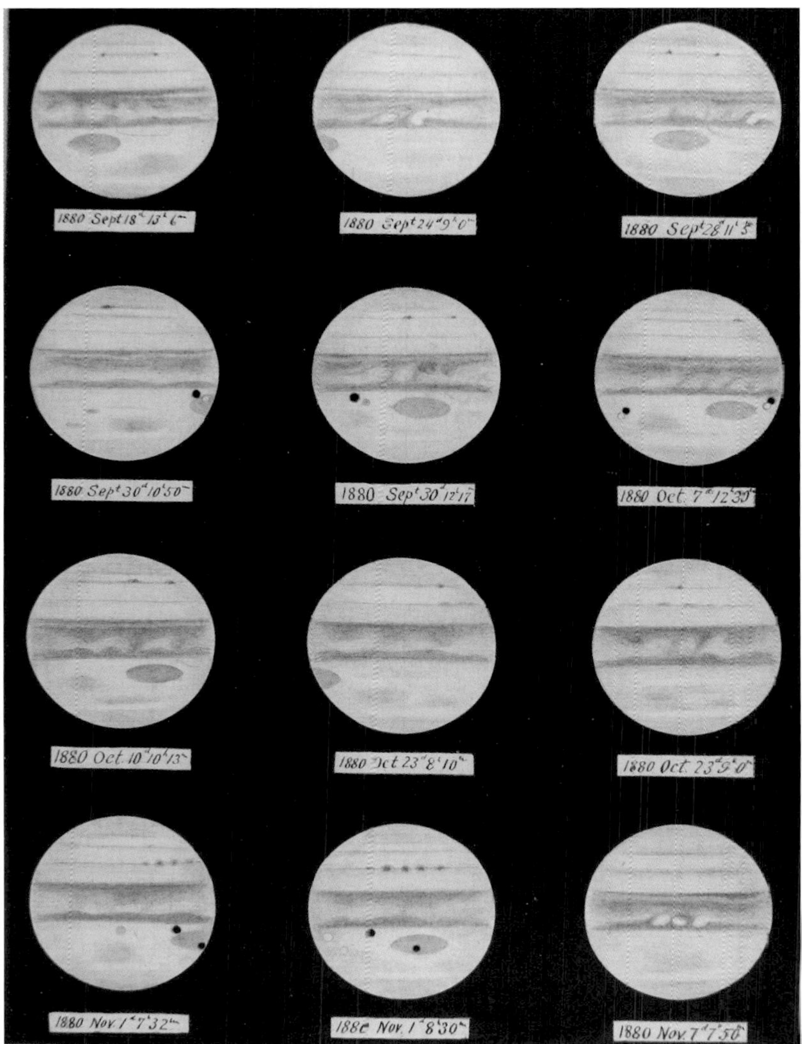

Barnard's drawings of Jupiter, from 1880, made with his 10-cm refractor in Nashville, Tennessee.

several years hovered near the brink of invisibility. It seemed the great show might be over. However, a revival took place in 1891. Since then the Great Red Spot has been prominent at certain times – for example, during most of the period 1961–75, in 1989–90 and again in 2013–14 – and at other times, such as 1977, 1991 and 2011–12, virtually invisible, though, as noted above, even then its position has always been marked by the notch-like Red Spot Hollow in the SEB.

In 1879–80 the Great Red Spot spanned nearly 35° of Jovian longitude. Ever since, it has been shrinking. By the time of the Voyager 1 and 2 flybys of 1979–80, its length had decreased to 21°. It had by then so fallen off from its prime that only one Earth would fit in it rather than three. By 2013–14 it had further shrunk to a mere 13.6°, giving it a length of only 15,900 km.

An archetypal example of other vortex features on Jupiter, and distinguished from them only by its size and longevity, the Great Red Spot rotates like a terrestrial storm, spun round and round as if it

Top: More of Nathaniel Green's superb pastel sketches of Jupiter, February 1883.

Bottom left: Instead of the warmer, yellowish and tawny colours dominant on the planet a year or two before, this drawing made with Green's new 46-cm reflector on 5 January 1882, shows a whiter Jupiter, with the Red Spot narrower and less reddish than it had been in 1879–80.

Bottom right: Jupiter, 9 July 1889, observed by James E. Keeler with the 30.5-cm refractor of the Lick Observatory, Mount Hamilton, California.

Even the most skilful observers differ significantly in how they see planetary detail – an effect known as the 'personal equation'. Compare Barnard's drawing from 1 November 1880 (last row, middle) with this drawing by the astronomer and astronomical artist E. L. Trouvelot, made at almost the exact same moment, from Cambridge, Massachusetts, with an 18-cm refractor. The orientations are different, as Barnard's drawing shows the planet oriented with north at the top while Trouvelot's shows it with south at the top – the latter being the usual telescopic orientation.

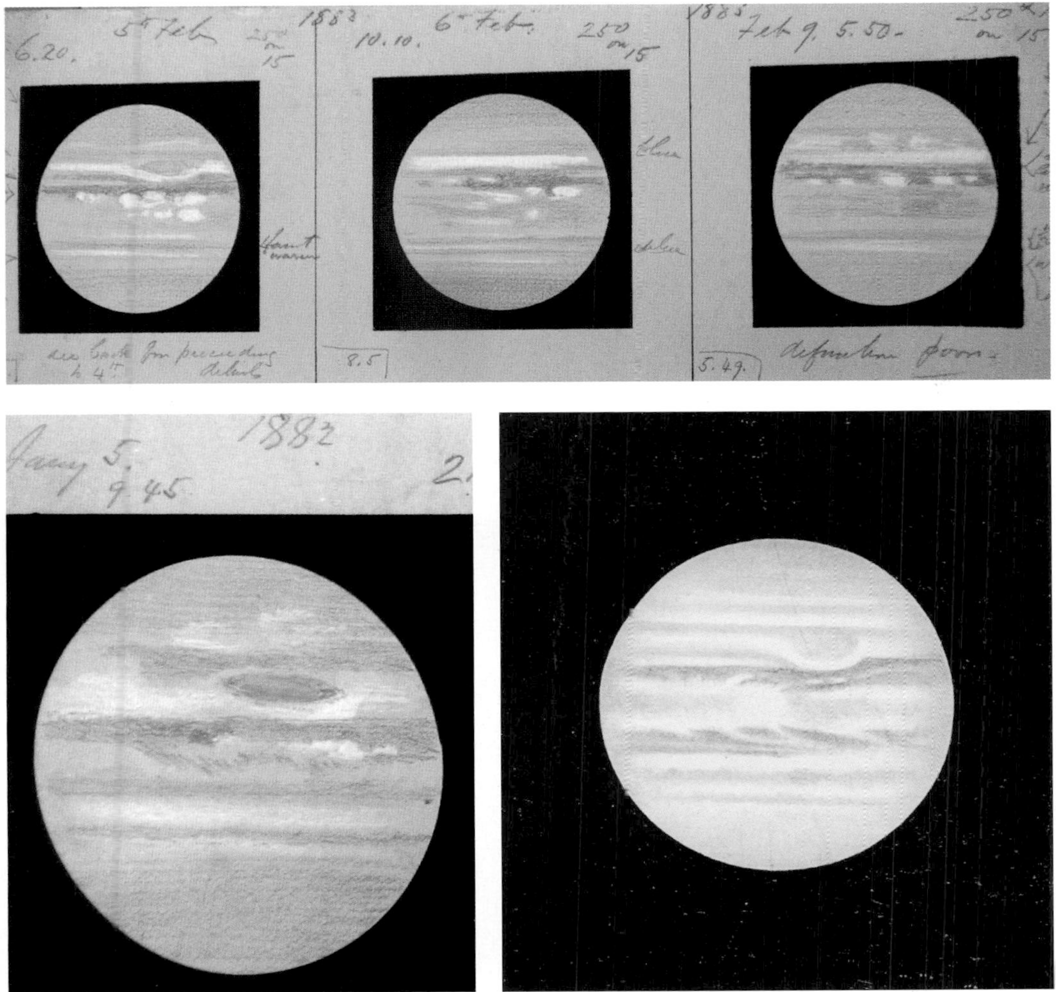

were a ball bearing between oppositely streaming wind jets lying north and south of it. The direction of spin is anti-clockwise, as would be expected for a high-pressure storm in the southern hemisphere of a prograde rotating planet. As smaller spots encounter the Great Red Spot, some are deflected away; others make multiple loops round it until eventually they are swallowed up in its maw. Their inward spirals are fascinating to watch, and it is both diverting

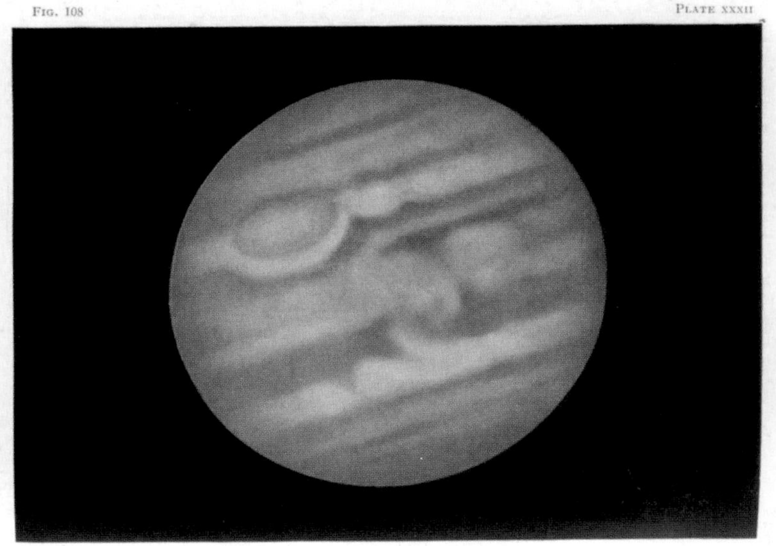

Jupiter, 9 June 1899, by Scriven Bolton. Bolton, a wealthy oil merchant, was unmarried, and lived with his widowed mother and sisters in Bramley, Leeds, where in a field adjoining his home he provided himself with a well-equipped observatory, which included an 46-cm reflector. He is best remembered today for his astronomical drawings and pioneering space art – and is seen as a forerunner of Lucien Rudaux and Chesley Bonestell. A polymath, he also composed several musical pieces for private performance.

Jupiter, 28 April 2016. A CCD view of the Great Red Spot, showing it to be much shrunken from its cigar-shaped form of the 1880s. This image, oriented with south to the top, was obtained by the Belgian amateur Leo Aerts, using a 25-cm f/30 Cassegrain and an ASI 120MM-S webcam, at 21.09 UT on 28 April 2016. The third satellite, Callisto, appears on the left, and is followed by its shadow projected just south of the North Temperate Belt.

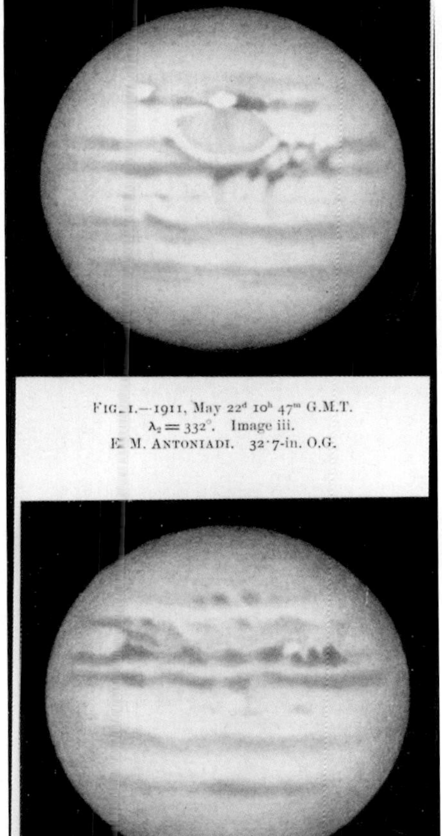

Antoniadi's drawings of Jupiter, made with the 83-cm refractor of the Meudon Observatory, near Paris, give a good idea of the complexities of Jovian detail as revealed by powerful instruments. At this time, there was much more detail in the southern hemisphere (at the top of these drawings, which are oriented to the telescopic view) than in the northern. In the drawing of 22 May 1911, the Great Red Spot is near the central meridian. In that of 12 June 1911, the whitish area is followed by a dark area in the South Tropical Disturbance, which had first appeared in 1901.

Jupiter, 10 August 1995, as it appeared with the 91-cm refractor of the Lick Observatory, x588. Drawing by David Graham based on observations by Graham and William Sheehan. The amount of detail visible in a large instrument like this, in superlative seeing, is mind-boggling. Note that the Great Red Spot has significantly shrunken from its glory days in the late 19th and early 20th centuries, and appears turbulent and highly disturbed. To the left of the Great Red Spot, on the South Equatorial Belt, are three white ovals which were then in the process of merging. Note the festoons extending from the North Equatorial Belt south into the Equatorial Zone. This drawing represents one of the parting efforts of the pre-CCD era.

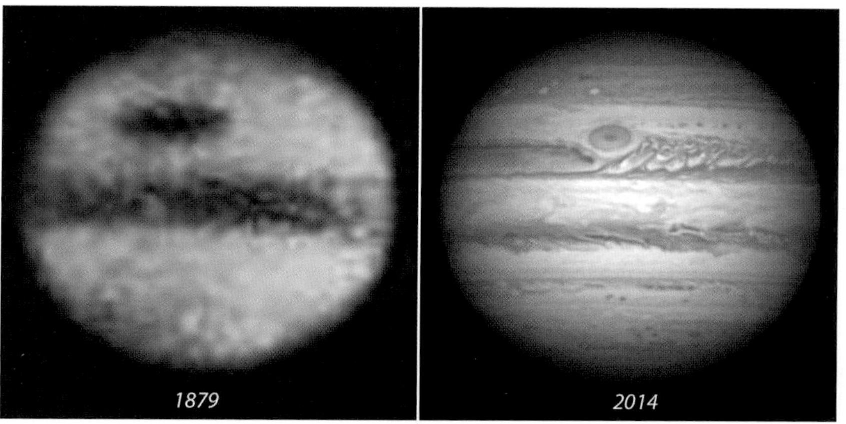
1879
2014

Jupiter's Great Red Spot as seen today compared to its appearance in one of the earliest (and hence grainiest) photographs of Jupiter, by the British amateur Andrew Ainslie Common, in its heyday more than 100 years before. Left: Frontispiece of Agnes Clerke, *A Popular History of Astronomy during the Nineteenth Century* (1908); right: NASA image, 2014.

and useful to make successive timings of their periods of rotation round the Great Red Spot.

The Great Red Spot's rotation period has varied over time. At the time of the Voyagers (1979–80), the period was six to eight days, corresponding to a mean wind velocity round the rim of 120 m/s (270 mph). By 2013–14 – as the Spot continued to shrink – it had decreased to 3.6 days, with a mean wind velocity of 144 m/s. The record mean wind velocity in the Great Red Spot to date is 165 m/s, observed by the Galileo orbiter in 2000. The changes in the Great Red Spot's rotation period are believed to occur as the vortex gains rotational energy by feeding on smaller spots swept along by the northern jet stream of the adjacent South Temperate Belt (STB).

There are certainly similar storm-like features in Jupiter's mid-latitudes, but there is really nothing comparable to the Great Red Spot. The lyrics of the old Cole Porter song 'You're the Top!' might well have included a reference to it; it is, among Jovian features, by far the most iconic – it is the Colosseum, and the Louvre museum. However, despite its long endurance – it is well into its second century, since it reached its maximum development around 1880 – it is a far cry from what it once was. Indeed, when one of us (W. S.) first observed it 1965 it could easily be seen in a 60-mm refractor with a magnification of only 35x, but in recent years it has not always been easy to see in even 15- to 20-cm reflectors. Anyone seeing it should savour the experience, for we do not know how much longer it will last. Like old soldiers, it seems Great Red Spots never die; they just fade away.

The Nature of the Great Red Spot: Early Speculations

When the Great Red Spot first came to widespread attention in 1878, no one quite knew what to make of it, and some interesting theories were put forward to account for it. The French astronomy popularizer Camille Flammarion, who had his own private observatory at Juvisy,

Juno image, taken from orbit round Jupiter in 2017, showing complex structure within the Great Red Spot.

Jupiter as seen from one of its more distant moons. The white circle on the right limb is Ganymede, with its shadow next to it; the grey disc to the left of centre is Callisto, which because of its low albedo appears very dark when seen against the bright background of the planet's surface. The Great Red Spot is near the centre. From T.E.R. Phillips, ed., *Hutchinson's Splendour of the Heavens*, vol. I (1925).

outside Paris, thought it might be a Jovian continent in formation. More plausibly, the well-known amateur astronomer William Frederick Denning, who was an accountant by profession but an astronomer by preference – today he is best remembered for his studies of meteor shower radiants, but he was also an indefatigable observer of the planets, using an excellent 30-cm reflector at Bristol – regarded it as possibly a gaping rent in the clouds through which were visible the dense dark vapours of its lower strata, and perhaps even the surface itself.

Describing photographs of Jupiter obtained at his Flagstaff observatory in 1909, Percival Lowell offered his own views about the nature of the Great Red Spot. Noting the remarkable longevity of the Spot or, what was appearing at the time, the Great Red Spot Hollow, he wrote:

> [The Hollow] is distinctly traceable in the drawings of Sir William Huggins made in 1859–1860, which he kindly sent me . . . Here then we have evidence of a feature which in its general outline has been stable and persistent for fifty years, a marvelous length of time for a cloud form as we know such things to continue to exist. This alone would suffice to demonstrate that Jupiter's meteorological conditions owe nothing to the Sun. That this cradle then became the centre of a vast ruddy mass, which after a time disappeared to leave it in its former condition, indicated it as the seat of some violent outburst from below, over which the cloud veil, rent at the height of the explosion, settled down again, covering the furnace from sight. That the feature is so permanent hints at a certain plasticity as opposed to what we call gaseous in the constitution, and likens it to those permanent seats of disturbance with whose vents we are familiar on Earth. A volcano in embryo is what its behavior points it out to be. It probably forms a connecting link between volcanoes on the one hand and sunspots on the other.[6]

In other writings Lowell echoed Flammarion's idea of a continent in formation, referring to the Great Red Spot colourfully as a 'sort of baby elephant of a volcano, or geyser, occurring as befits its youth in fluid, not solid, conditions, but fairly permanent, nevertheless – a bit of kindergarten geology'.[7] Though such ideas may seem naive today, they were not entirely wrong. The Great Red Spot is a volcano of a sort, though of a distinctively Jovian kind; it towers up, as we shall see, above an eruption or bubbling up of a column of hot material from below.

An Amateur's Planet

After the rise into prominence of the Great Red Spot, Jupiter, hitherto the province of the occasional devotee, began at last to receive the attention its bulk and dominant role in the solar system deserved. For the better part of a century amateurs played the leading role in these studies, and continue to make important contributions right up to the present time. Indeed, the astronomical literature bears abundant testimony to the glorious work of amateurs, and yet so fleeting is fame that most of those who added so much to our understanding of the Giant Planet are now but little known. Nathaniel Green, Arthur Stanley Williams, William Frederick Denning, Theodore E. R. Phillips, Bertrand M. Peek and F. J. Hargreaves are but names. Nevertheless, we owe them a great debt, for the record they kept of the changes in the planet's cloud features has proved indispensable to our understanding of its long-term meteorology and still informs us to this day.

At first blush the study of Jupiter with even a modest instrument seems daunting. The sheer quantity of the details and the drama of the sweeping changes taking place in them seem impossible for the human eye-brain-hand system to effectively capture. Moreover, the would-be student of the planet who returns to the old memoirs of the Jupiter Section of the BAA, for instance, hoping to draw inspiration and perhaps pointers as to technique, is faced with a forbidding aspect: though the planet is a visual feast, and presents more interesting detail than any object other than the Moon, the records have long been dominated by accountant-like tables of timings of spots across central meridians and derived rotation periods, and charts of drifts in longitude. On opening the pages of these memoirs, one feels in the presence of a chronicle, not a story. It is only too tempting to turn away in disgust, as from the lessons of a drab schoolmaster.

Every subject takes on the colouring of those who are drawn to it. The mapping of the Moon has drawn more than its share of obsessives, keen to capture every small detail (and, in the Moon's case, for a long time missing the forest for the trees). Jupiter too has found its particular 'type'.

This 'type' was set by Arthur Stanley Williams, by profession a British solicitor and one of the legendary figures in the study of Jupiter. A native of Brighton, he already was observing Jupiter as a teen, won over (as so many of the time were) by the Great Red Spot's dramatic rise to prominence. However, his serious study commenced only in 1886–7. He was greatly stimulated by the founding of the BAA in 1890 and soon became a leading member, despite having instrumentation that was modest by any standard – for most of his observations he used only a 16.5-cm reflector which, though mounted equatorially, was without a clock drive. It was true of Williams, as C. P. Snow once said of Ernest Rutherford, that he worked with bizarrely simple apparatus, but carried the use of such apparatus as far as it would go. Only in the last year of his life did Williams finally acquire a better instrument, a 23-cm reflector.

Though possessed of a keen eye, Williams was a notoriously maladept draughtsman. In contrast to Nathaniel Green, another leading Jupiter observer of that era and a portrait painter by profession whose drawings of Jupiter, in colour, are works of art, Williams produced drawings that can only be described as pragmatic; they

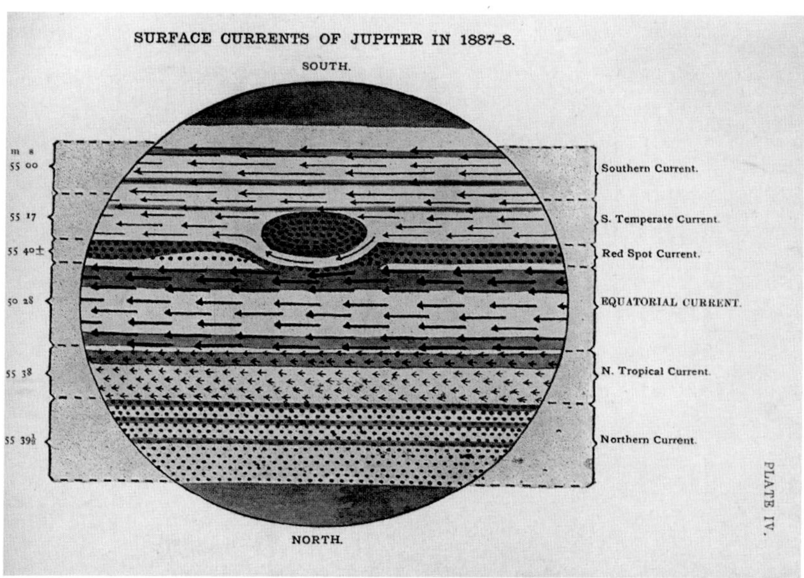

Arthur Stanley Williams's wind currents, 1887–8.

were justly characterized by Bertrand M. Peek in a BAA Memoir of 1938 as 'diagrammatic rather than pictorial'.[8] Here, however, necessity proved the mother of invention. Instead of going against the grain and trying to record the exact form of the various features, Williams took to estimating visually, to the nearest minute, the time that each feature transited Jupiter's central meridian in order to investigate the existence of wind currents on the planet. (It was thus that the memoirs of the Jupiter Section of the BAA came to assume their accounts-like aspect.) By following a feature through several transits and plotting its observed transit times on graphs, he was able to work out the rotation periods of individual features and determine what he called the 'independent drift' in different latitudes of the planet – in effect, using each feature as a kind of 'windsock' to probe the Jovian winds.

Williams attempted as far as possible to be atheoretical. He cautioned that when he used the term 'surface material', he meant to suggest no opinion about its nature; 'by the term . . . is meant [only] the material at the *visible surface* of Jupiter, without reference to the question whether such material is actually at the solid, or liquid, surface of the planet, if, indeed, the latter can be said to possess such a thing as a solid or liquid "surface"'.[9] His caution was admirable. Less creditably, he introduced the terms 'canal' and 'channel' into the Jovian nomenclature – terms which were to create so much difficulty with respect to Mars. Fortunately, these terms never caught on and the features he described as 'canals' were later more aptly referred to by the British amateur Scriven Bolton as 'wisps'. On the other hand, one sees the term 'barge' used from time to time, a term which not only exactly describes the Jovian feature to which it is applied but recalls Williams's preferred residence – a house boat – in later years.

The monumental labour of taking timings and performing reductions of data was, in the pre-calculator and pre-computer era, 'simply enormous', as Williams affirmed, 'though the real meaning of it can only be properly appreciated by those who have made and discussed, say, 1,000 determinations of the longitudes of different markings'.[10]

'Wisps' on Jupiter, as sketched by Percival Lowell in 1907.

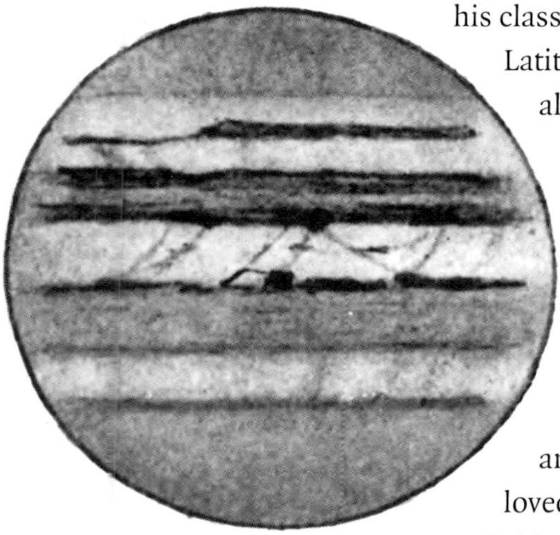

By 1896 Williams had progressed to the point where he had identified nine distinct currents on Jupiter, as described in his classic paper 'On the Drift of Material in Different Latitudes of Jupiter'.[11] One might imagine that, along with his solicitor's work, keeping tabs on Jupiter would have left Williams time for little else. One pictures a rather lonely and determined man soldiering on in pursuit of his peculiar predilection. (Like many of the great amateur observers, including W. F. Denning and Percy Molesworth, Williams never married.) In fact, he was a man of parts, and had other equally absorbing interests. He loved the sea. On his retirement, Williams lived at St Mawes, Cornwall, on a barge named the *Queen*, with his observatory on shore nearby, while in 1920 he won the Challenge Cup – then the highest award given in world cruising – for a single-handed cruise of 2,000 km from Falmouth to Vigo and back. At his death in 1938 his body, according to his express wishes, was taken by steamer to a point off Falmouth a few miles from land, and committed to the sea in the presence of just two friends.

Williams's paper on the circulation of Jupiter inspired a number of like-minded members of the BAA to cultivate the same line of work. One was Denning, who was introduced above. An accountant by profession, he would have been well suited to taking meticulous records of the kind required to monitor the weather on a planet like Jupiter. Of a retiring nature, like Williams, he gave up a promising career as a cricketer to pursue astronomy. In later years he lived alone in Bristol with his manservant and managed to accomplish a great deal despite limited means and ill health. A note by the British historian of astronomy Richard Baum helps to bring him back to life:

> Years ago the late W. H. Steavenson, a very well-known English amateur, visited Denning, to discover a lonely old man sitting by his fireside. James Muirden, author of the *Amateur Astronomer's Handbook*, told me that tale, and added that Muirden's father lived a few streets away from Denning and could well remember how the street urchins used to catcall and abuse him as he made his way home.[12]

Another noted figure of the late nineteenth and early twentieth century was Major Percy Molesworth, who was born in Ceylon (now Sri Lanka) but educated in England preparatory to a military career. He served in England and Hong Kong but was then transferred back to Ceylon. Installed as an army engineer at Trincomalee, he used a large reflector set up in his garden in that near-equatorial location. Though he was sometimes away on active service for months at a time, in between he seems to have had ample leisure and, not having the care of a wife or family, was an indefatigable observer of the Moon and planets. Jupiter of course came in for its fair share of attention. In 1900 alone Molesworth made 6,758 timings of features crossing the central meridian, then devoted months to slavishly deriving from these timings the features' rotation periods. He attained the rank of major by the time of his retirement in 1906. Unfortunately, he did not ease up on the demands he placed upon himself; he continued to work too hard and in the torrid climate this injured his health. He died of dysentery in 1908, at the age of only 41.

One of the features Molesworth recorded on Jupiter has become the stuff of legend. This was referred to as the 'South Tropical Disturbance' and after the Great Red Spot ranks among the most celebrated features ever observed on the planet. Molesworth noted that in 1900 the South Tropical Zone had been almost the brightest of the zones, 'distinct and brilliant, brightening to a milky white where it borders the S[outh] equatorial belt'.[13] However, by February 1901 its appearance had changed: now a series of dark humps appeared on the south edge of the SEB, and by May one of these humps had developed into a dark shading reaching across the South Tropical Zone all the way to the South Tropical Belt. This dusky part of the belt was quite distinctive and became known as the South Tropical Disturbance. Because it had a slightly shorter period than the Great Red Spot, the South Tropical Disturbance caught up with the Great Red Spot from time to time; however, instead of overlapping with it, it seemed to 'leap' across, and re-formed completely on the other side. The South Tropical Disturbance was also famous for blocking the passage of smaller spots, which, as they approached, were deflected back in the direction from whence they came. The South Tropical Disturbance was long-lived and lasted until 1940. Since then there have been similar disturbances, including one between 1955 and 1958 and another between 1978 and 1983 – the latter, fortuitously, occurred during the Voyager flybys, and could be studied in detail. It is now

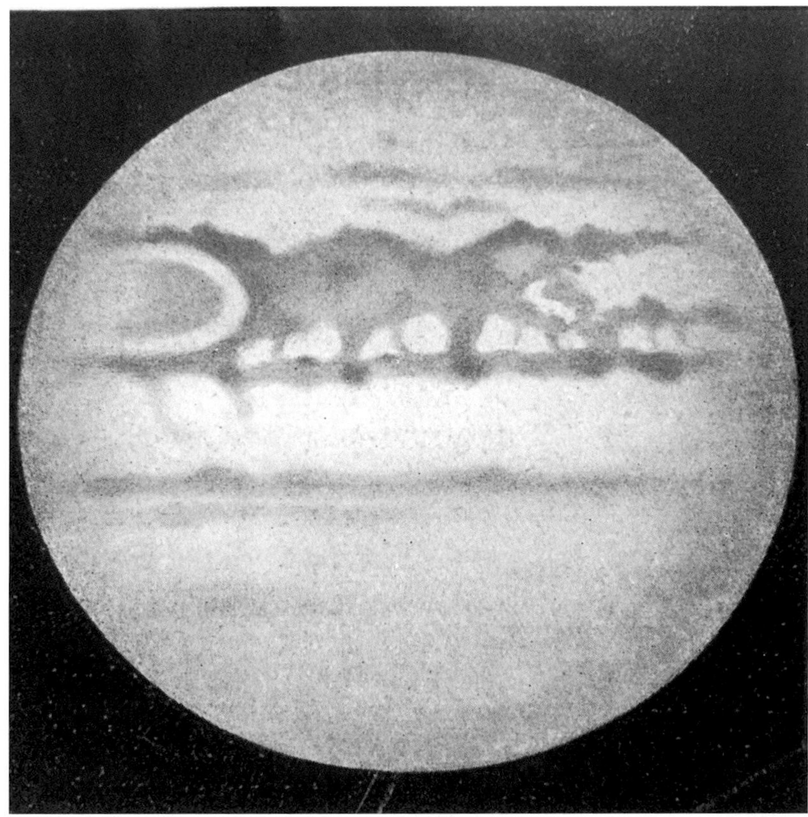

Jupiter according to another skilful artist-astronomer, E. M. Antoniadi, 21 May 1901, using the 24-cm refractor at Camille Flammarion's observatory at Juvisy, near Paris. This shows the Red Spot nestled in its 'hollow'. The mass of dark matter to the right is part of the South Tropical Disturbance, which appeared at this time. This drawing was made near the end of Antoniadi's tenure as Flammarion's assistant.

known that a South Tropical Disturbance connects the retrograding South Equatorial Belt south jet stream to the prograding South Tropical Belt north jet stream, giving rise to what is known as the Circulating Current.[14]

Meanwhile, in 1901 – the first year of the new century – an enormous new talent first trod upon the stage of Jovian studies. This was the Reverend Theodore Evelyn Reece Phillips, who became director of the BAA Jupiter Section, a position in which he continued to serve until 1934. In person Phillips was a tall, thin, rather sour-looking figure of a type not unfamiliar among the Anglican clergy of those days. Appearances were deceiving: in reality he was mild-mannered and self-effacing as well as, of course, highly

Jupiter, 12 August 1903. These two drawings, made only two hours apart by the famous Jupiter observer T.E.R. Phillips, illustrate the rapid rotation of the planet. The darkish area shown near the right limb in the drawing on the left and near the centre in that on the right is the famous South Tropical Disturbance.

A characteristic view of the South Tropical Disturbance, as it looked later in its development. This drawing was made by Phillips on 19 December 1917, with a 30-cm reflector.

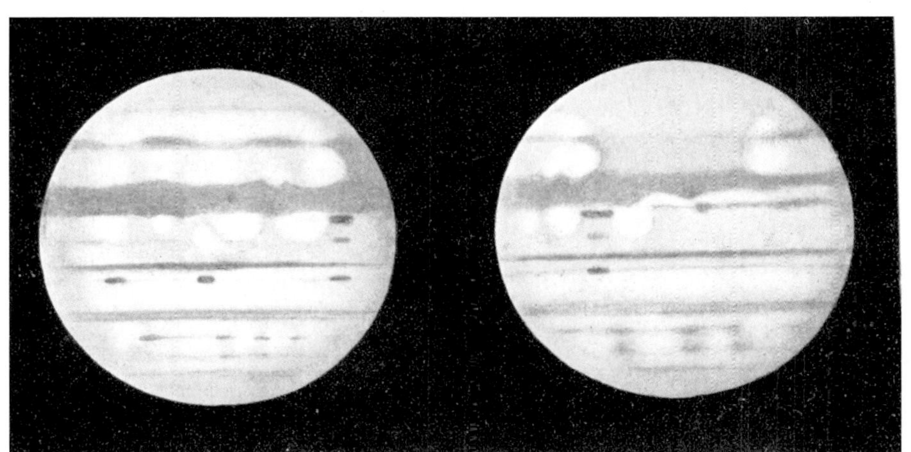

1917 Dec 19d 6h 45m G.M.T

conscientious and possessed of limitless drive. In 1914 he began to combine the transit observations – the timings of features across Jupiter's central meridian – by all the better observers to plot more reliable drift charts. Two years later, he became vicar at Headley near Epsom, Surrey, where he set up a private observatory, equipped with a 20-cm refractor and the same 50-cm reflector formerly used by the great astronomer-artist Nathaniel Green.

The legendary Jupiter observer Theodore Evelyn Reece Phillips (1868–1942). This portrait hangs in the hall outside the office of the British Astronomical Association.

Since Phillips, other leading Jupiter observers have included Bertrand M. Peek, an outstanding mathematics student (and tennis player) at Cambridge who served with the Hampshire regiment during the First World War and later became a school headmaster at Solihull, near Birmingham, and F. J. Hargreaves, a patent agent at Coulsdon who became a professional telescope-maker. They were quintessentially British; one pictures the three of them – Phillips in the black suit and stiff white collar of the clergy, the others in tweed – on the lawn having afternoon tea. They kept remarkable records of the meteorology of the planet – a feat that becomes even more remarkable in light of the notorious

Phillips's private observatory at Headley (near Epsom, Surrey), where he served for many years as Vicar. Phillips is standing near the centre of this image, to the left of the man looking out the window of the small observatory. The large instrument in the foreground is the 50-cm reflector that had been used by Nathaniel Green in the late 19th century.

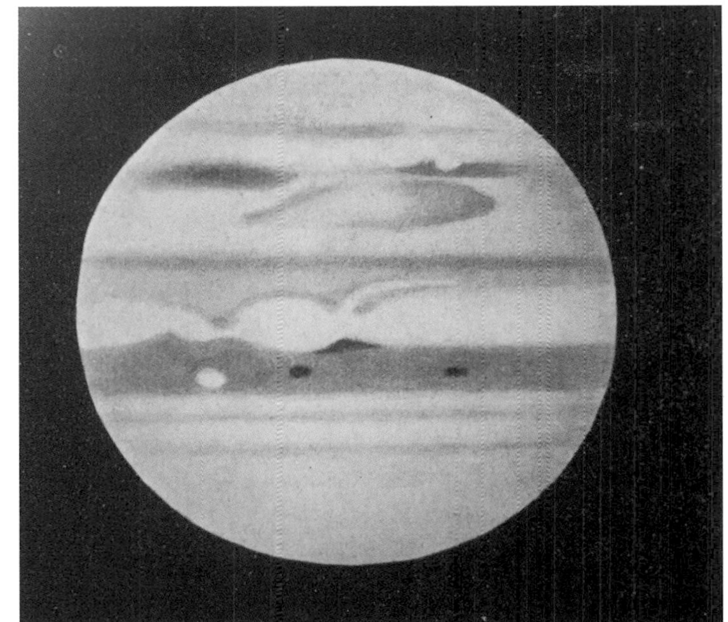

Jupiter, 29 November 1919. Drawing by Phillips, using a 30-cm reflector. This drawing shows the complete disappearance of the southern portion of the South Equatorial Belt and the Red Spot Hollow. The peculiar elongated region in the South Tropical Zone is the Great Red Spot, which became more prominent as its surroundings faded away. Note the broad North Equatorial Belt with barges and a 'porthole'.

Jupiter, 30 September 1926. Drawing by Phillips, showing the Red Spot at the left of the disc, hovering above a faded South Equatorial Belt. The North Equatorial Belt is disturbed. This is the appearance of the planet prior to the great South Equatorial Belt revival that began in August 1928.

vicissitudes of the British weather – for forty years. Peek summed up the work of this team and other BAA observers in the *Memoirs* of the BAA in 1940:

> There seems to be no other published series of planetary observations that is in any way comparable with them, constituting as they do, from 1901 at the latest onwards, a continuous record of the appearance and motion of the surface features during the whole time that Jupiter was in a position to be observed . . . Any student of Jovian meteorology must necessarily turn to these *Memoirs* for the data on which to base his researches, for nowhere else will he find what he requires.[15]

The years when Phillips, Hargreaves and Peek were active marked the heyday of the BAA's Jupiter Section. The planet was, to an extent that it will never be again, the Amateurs' Planet. Their heroic effort did not long survive the Second World War, however. The nerve-centre of Jupiter studies, Phillips's Headley, lying within the Home Counties, sustained heavy damage. Phillips's own church was bombarded, and a V1 'buzz bomb' even landed in the Rectory. It must have been nerve-wracking to have it there, but fortunately it failed to go off. Perhaps hastened on by the events of that terrible time, Phillips's health began to fail. He died in 1942. Peek, who had succeeded him as director of the BAA Jupiter Section in 1934, dismantled his own observatory in 1947 as his own health began to fail. He remained in the directorship of the BAA Jupiter Section and continued to produce memoirs of the section for two more years, and then began to write a book, *The Planet Jupiter*. Published in 1958, it remained the definitive account of the planet until the spacecraft era, and even now is not without value. (T. H. referred to it in his doctoral dissertation.)

After Peek's retirement, and with Jupiter's motion carrying it to a more southerly location in British skies, the BAA Section retained few active members. Instead, in the years between 1949 and 1963, the best Jupiter observations were made by an American, Elmer J. Reese. After serving with the United States Army Ordnance during the Second World War, he returned to his hometown of Uniontown, Pennsylvania, where he worked in his family's grocery store by day and observed at night with a 15-cm home-built reflector. This instrument was even smaller (by 1 cm) than the one Williams had used. The fact that such an instrument, in the right hands, could be used to do valuable work shows just how much scope for research Jupiter affords. Reese became known for the accuracy of his drawings and transit timings, and from 1947 served as the first Jupiter Section Recorder of the newly founded Association of Lunar and Planetary Observers (ALPO), in which position he was responsible for drawing up most of the section's reports. (For some reason, while the heads of the BAA observing sections are known as directors, those of the ALPO observing sections are known as recorders.)

Reese himself contributed many observations of the South Temperate White Ovals, which developed in the early 1940s. He also investigated the great SEB disturbances and revivals (returns of the Belt from obscurity to prominence). Despite his lack of professional credentials, in 1963 he was invited to join the planetary research group at New Mexico State University led by Clyde Tombaugh and Bradford Smith and collaborated in one of the first professional programmes to monitor Jupiter, regularly photographing the planet with a 61-cm reflector and laying the groundwork for the Voyager spacecraft missions. By the time he died in 2010, at the age of 91, he had had the satisfaction of seeing the successful completion of the two Pioneer missions to Jupiter, the two Voyager missions and the Galileo orbiter. Since then, several more probes have visited Jupiter on the way to other destinations – Cassini on the way to Saturn, and New Horizons on the way to Pluto – while yet another orbiter, Juno, entered orbit around Jupiter on 4 July 2016.

Under an Ever-changing Jovian Sky

Though Arthur Stanley Williams will always be remembered for his discovery of the nine main currents of Jupiter's wind circulation, detailed knowledge has largely been the work of his successors – and of teams studying data returned by spacecraft. No fewer than nineteen currents have now been recognized between latitudes 60° North and South (many having first been recorded by the Voyager space probes). The wind currents seem to be remarkably constant over time. There were, for instance, no differences noted between Voyager and Hubble Space Telescope measures made at the same Jovian season twelve years (one Jovian year) apart.

To recap what was discussed above, the basic scheme of Jovian meteorology includes the fact that the brighter zones – where the atmosphere is ascending – appear to be high-pressure, while the dark belts, areas of atmospheric descent, are low-pressure. A meteorologist trained on the Earth will be somewhat disoriented: on the ground, high pressure is associated with sinking air and low pressure with rising air. Once again, one has to keep in mind that on Jupiter they do things differently: there is no ground and the pressure is reckoned at the cloud tops.

The belts and zones also are broadly correlated with the 'slow' currents discovered by Williams and well known to other visual observers of the planet. The currents are, however, according to BAA Jupiter Section Director John H. Rogers,

> more invariant than the belts, with modest wind speeds defined by the motions of large ovals and other features. Each of these 'slow currents' coincides with one belt/zone pair.

> The slow currents coexist with a pattern of much faster east-west currents, called jets or jet streams, which seem to form the fundamental framework of the belt/zone pattern. The jet streams sometimes show outbreaks of small spots, which disclose their locations to visual observers. They were revealed in elaborate and complex detail only when the Voyagers imaged the small-scale cloud texture and revealed a continuous zigzag pattern of winds, alternating between eastward and westward jet streams. The latter correspond to the long-term edges of the belts and zones. While the small-scale features move in a flow pattern, larger spots – actually vortices that 'roll' between the jet streams – interrupt it.[16]

Though a great number of rotation periods for various cloud features were amassed by observers of the planet, up until the 1950s, the reference frame – the Jovian equivalent of the Greenwich Prime Meridian – had to be defined somewhat arbitrarily for the simple reason that no visible solid surface was identified.

For convenience, the planet had long been subdivided into two main latitude regimes. Longitudes of spots within about 9° either side of the equator – those being rushed along by the Great Equatorial Current – were calculated according to System I, with a period of 9 hours, 50 minutes, 30.0 seconds. The longitudes of spots at higher latitudes were calculated according to System II, with a period of 9 hours, 55 minutes, 40.6 seconds. (This had been the mean rotation period of the Great Red Spot in 1890–91, when the scheme was first adopted.) This subdivision of the planet into Systems I and II allows for a reasonably good idea of the average relative drift of features lying within the two regions, though obviously individual spots within each zone will have their own independent motions.

In 1955 the rotation of the core of the planet was at long last established by radio astronomers, who recorded intense, intermittent bursts of radio-frequency energy emanating from deep within the interior of the planet. This work defined yet a third system of longitude: System III, with a period of 9 hours, 55 minutes, 29.4 seconds, which was tied not to the planet's visible cloud features but to the planet's core.

The discovery of the 'true' rotation period of the core of the planet marked a great leap forward. Now, for the first time, detailed modelling of the planet's circulation was possible. According to John H. Rogers:

> The currents recorded by visual observers fall into three categories. First, there is the great equatorial current: the entire equatorial region between about 10° North and 10° South progrades (i.e., carries features in an eastward direction relative to the core) at 7–8° per day (approximately 100 m/s) relative to System II, and intermediate speeds are sometimes recorded along its edges. Second, there are the nine slow currents, which govern most of the visible features outside the equatorial region, and have speeds of no more than 1° per day (<10 m/s). Third, there are the jet streams on the edges of certain belts, only observed during infrequent outbreaks of small dark spots, which have speeds of several degrees per day (30–170 m/s). These now are known to be permanent.[17]

Based on Voyager results (and confirmed by Galileo), the highest wind speeds on the planet are found in two narrow jet streams located in the Equatorial Zone (EZ), at 6° to 7° north and south of the equator, which reach 170 m/s. In the northern hemisphere, as one travels further away from the EZ, the winds decrease and then reverse direction; at round 15° North they become westward instead of eastward, at 25 to 70 m/s. At about 20° North latitude there is another narrow eastward jet – the wind speeds here are more than 150 m/s – while still other eastward jets are found at latitudes 32°

and 38° North. In still higher latitudes, all the way to the pole, the winds fall off steadily. In the southern hemisphere, the locations and speeds of the jets are somewhat different from those in the northern hemisphere. For instance, the westward jet there is located at 17.5° South.

SIX

A Bewildering Phantasmagoria: Jovian Meteorology

As noted earlier, because of its unusual persistence, the Great Red Spot was long thought to be attached to an underlying surface feature of some kind (an erupting volcano, perhaps, as Percival Lowell suggested). However, this idea cannot be reconciled with the fact that the Spot drifts round in longitude while also undergoing occasional marked accelerations and decelerations. Thus it was moving more rapidly between 1920 and 1940 than it had in recent years, and as was first noticed by the British amateur Jupiter specialist Bertrand Peek, the Red Spot seems to darken in colour when it slows down. Realizing that these peregrinations prove that it cannot be anchored to the surface, Peek supposed that the Spot might be a solid body floating like an egg in salt water. It was a charming idea, but has proved to be completely wrong. In fact, the Great Red Spot is a purely atmospheric phenomenon, albeit a most unusual one.

What is It?

We now know that the Great Red Spot is actually a towering high-pressure area whose cloud tops loom several kilometres above, and at a cooler temperature than, the other Jovian clouds. The Great Spot may go so far as to *transport* a great deal of heat from the interior of the planet to its high atmosphere (just as a hurricane extracts heat from the warm ocean). As suspected by the American amateur Elmer J. Reese from observations with a 15-cm reflector in 1949, and as finally proved in the 1970s by Reese, Bradford Smith and Gordon Solberg in photographic work at New Mexico State University, the Great Red Spot does have an internal rotation like a hurricane. (Remember that as a feature in the southern hemisphere, the Great Red Spot's rotation is in an anti-clockwise or anti-cyclonic direction.)

When Reese and Smith published their findings, the rotation period was six days, corresponding to wind speeds of about 100 m/s. The Great Red Spot vortex is trapped within a narrow zone of latitude – between 15.4° and 25.4° South – by two opposing jets, which are strongly deflected to either side of it to create the Great Red Spot's sharply sculptured oval contour. Smaller eddies approaching the Great Red Spot from the east tend to be swept round its north side, but as they move toward the west end of the spot some are pulled into the general circumvolving current and dump white ice into the spot to form a collar of white clouds. In the spot itself the darker interior is dominated by small convection cells, which can sometimes be glimpsed by visual observers or recorded by CCD imagers using large instruments.

Perhaps the most compelling of all the Great Red Spot's mysteries, with the possible exception of its longevity, is its colour. According to one long-held theory, the colouring might be produced by chemicals formed deep within the planet and dredged up to the upper cloud layers. Molecules containing sulphur or phosphorus, as well as organic molecules, were leading contenders, while lightning was hypothesized as a source of energy for molecular change. However, recent analysis by the Jet Propulsion Laboratory's Kevin Baines and his colleagues of data from the Cassini probe, which flew by Jupiter en route to Saturn in December 2000, shows that the reddish material in the Great Red Spot may be confined only to the upper layers of the clouds. Moreover, the reddish material turns out to be a close match with a concoction produced in the laboratory by ultraviolet irradiation of ammonia and acetylene gases, both known to be common in the hazy upper atmosphere of the giant planet. Instead of being a towering column of reddish cloud, or the eruption of a deep-seated volcano, the reddish material proves to be a mere superficial glaze; further down, the clouds are likely whitish or even greyish.

In the Great Red Spot ammonia ice is transported higher into the atmosphere than usual, where it is less shielded from the Sun's ultraviolet light, but in addition the high winds associated with the vortex motion trap reddish compounds, making escape difficult and deepening the colouring. The variable intensity of the Red Spot's colouring appears to be due to changes in the amount of white cloud it is 'ingesting'.

Indeed, since its heyday in the 1870s and '80s, when it appeared sausage- or cigar-shaped, the Great Red Spot has undergone marked changes. The most obvious is that it has become progressively less elongated over time, though even this has not been an entirely uniform process: between 1996 and 2015 it showed less tendency to shrink in a north–south direction, and assumed a rather rounder outline. Accordingly, it looked more like a rugby ball than an American football. Throughout this period, according to John Rogers, who has been tracking grey streaks and other features within the Spot, its internal circulation period has continued to decrease.[1]

Thus from the six-day period found by Reese and Smith in the early 1970s, the period has shortened to 4.5 days in 2006 and to only 4.0 days in 2012. The angular momentum of the spot has not changed, however, and so – by analogy to a skater pulling in her arms – the outer wind speeds within the spot have increased. In recent years, they have clocked in at from 270 to 430 to as much 680 km/h. By comparison, terrestrial hurricanes – for instance, the devastating Hurricane Katrina of 2005, which came ashore at New

Orleans with wind speeds on the order of 185 km/h – would seem almost a gentle breeze by Jovian standards. Jupiter is indeed, as Huygens called it, a windy planet.

Though the Great Red Spot has shrunk by 50 per cent of its original size, interacted with numerous other features and spots, changed its drift rate and, at least in recent years, shortened its internal rotation period, it has, oddly enough, consistently maintained an oscillation in longitude with a ninety-day period throughout the time it has been monitored.[2]

The White Ovals

The Great Red Spot is obviously the most long-lived of the Jovian atmospheric features. Its long-term dynamical stability appears to be partly owing to its position near the equator and the absence of 'ground' friction. However, a few other spots have also been quite long-lived. Of these, the best examples are the so-called White Ovals.

Voyager 2 image of Jupiter, obtained during the spacecraft's 'observatory phase', in June 1979. This image, taken from a distance of 24 million km, shows the Great Red Spot, followed by a region of chaotic and turbulent clouds, and one of the White Ovals, which is followed by similar turbulent whitish clouds in its wake.

When the Hubble Space Telescope obtained this image in 1998, the long-enduring White Ovals – A-B, C-D and E-F – monitored from the Earth since the 1940s were grouped closely together behind the Great Red Spot and about to begin a series of interactions that unexpectedly led to their merging in 2000.

Juno close-up of one of Jupiter's White Ovals.

The White Ovals are also anti-cyclonic systems and are closely associated with the South Temperate Belt (STB), which forms the southern boundary of the zone in which the Great Red Spot travels. Their history dates back to 1939 when, according to BAA Jupiter Section records, the STB was noted to be brighter than normal; however, it contained three brownish segments, given the rather perfunctory designations A-B, C-D and E-F. By the late 1940s, it became evident that as the segments became more and more spread out the lighter intervals between were developing into organized cloud systems: the White Ovals, F-A, B-C and D-E. Drifting eastward relative to the Great Red Spot, they caught up with and passed south of it about once every 2.6 years. Moreover, since each moved with a slightly different period, they approached or receded from one another without merging. Thus, B-C approached to within 50° of the longitude of F-A in 1975, then fell back again as F-A began to accelerate. For some years the White Ovals appeared to be shrinking and it was generally expected that, once they fell below a certain critical threshold of size, they would break up owing to atmospheric turbulence. However, after a series of interactions occurring between 1998 and 2000, B-C, D-E and F-A unexpectedly consolidated into a single large White Oval.

A close-up look at the so-called Red Spot, Junior, which formed when three long-lived White Ovals merged, bringing about a change in colour. This spectacular image was obtained by the Juno spacecraft, 11 December 2016, during its third flyby of Jupiter, and shows the view 16,600 km above Jupiter's cloud tops.

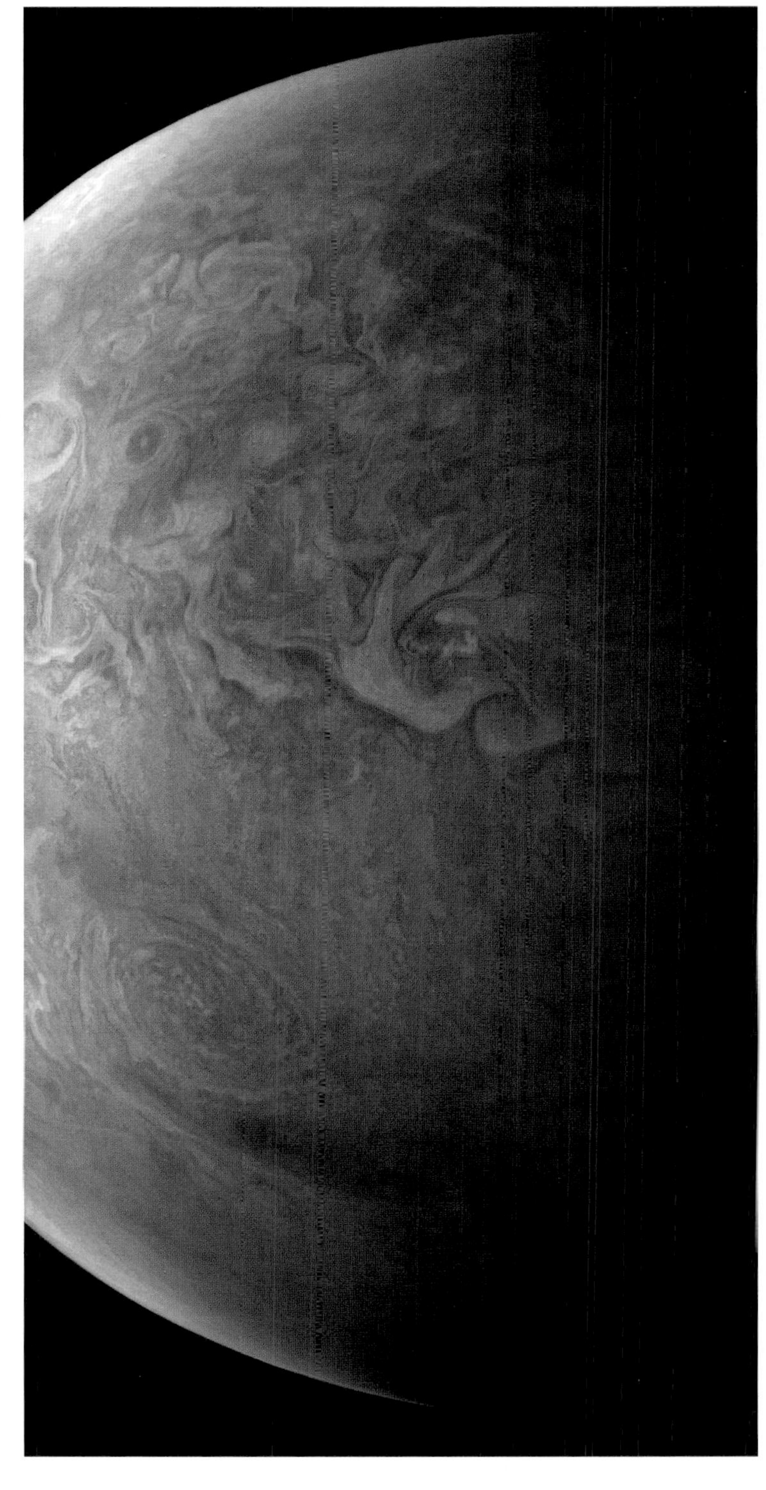

In February 2006 the White Oval B-A underwent a colour change to red, first noted by planetary imager Christopher Go, an amateur astronomer from Cebu, the Philippines. This event triggered keen interest among amateur and professional astronomers alike.

Presumably the change in colour was owing to the increased height and fiercer winds of the post-merger feature. Since then, the feature, which has been referred to as Red Spot, Junior, has usually been red, though at times it has appeared mostly white with only a tincture of reddish colour. Amateur and professional observations have shown that the motion of this anti-cyclonic feature is critically affected by South Temperate Belt (STB) segments impinging upon it. Clearly, these developments are suggestive, and it is not difficult to see in this sequence of events a plausible scenario for the development of the Great Red Spot itself.

In addition to the famous White Ovals, there are many smaller eddies, most of which come and go with great rapidity, and so are but the evanescent features of a Jovian summer's day. Some, however, are longer lasting, and can be followed from one Jovian apparition (the season when Jupiter is well enough placed to be accessible to Earth-based observers) to the next. These longer lasting features include a well-defined string of White Ovals in the South South Temperate Zone (SSTZ), and a long-lived white oval in the North Tropical Zone known as 'White Spot Z' (WSZ), which merged with another oval in 2013, after which the merged feature turned pale red. (Again, a sequence suggestive of what might have been the Great Red Spot's incarnation.) Yet another White Oval in the South Equatorial Belt (SEB) underwent a colour change to red in May 2009 but, only a few months later, it collided with the Great Red Spot, which apparently found it unpalatable and spat it out again; subsequently the disgorged spot lost its reddish colouring and relapsed into a White Oval. Such is the dog-eat-dog – survival of the fittest – the nasty, brutish and short life of a Jovian anti-cyclone.

The South Equatorial Belt Disturbances

The Equatorial Zone (EZ) marks the region of the powerful Equatorial Current – the region of System I – which hurries features along at a rate of 8° of longitude per day relative to features in other parts of the planet. The EZ is bounded on the south by the South Equatorial Belt (SEB) and on the north by the North Equatorial Belt (NEB). During most of the twentieth century and so far in the twenty-first, the NEB has been the most prominent belt on the planet and the scene of numerous dark projections (plumes, festoons or blue features) along its northern edge. However, the polarity was reversed in the late eighteenth and nineteenth century. Then the SEB was more prominent and displayed plumes or blue features along its south edge. The reversal took place in 1911.

The SEB undergoes a major cycle of fadings and subsequent revivals to former intensity. A review of the historical record shows that several such cycles apparently occurred in the nineteenth century, notably in 1869, 1873 and 1882. There is no evidence of cycles between 1882 and 1918, during which period the SEB remained the broadest and usually the darkest belt on the planet. A dramatic change took place in late 1918. The SEB's southern component suddenly began to fade from view, and the entire region became intensely white. This was followed, in December 1919, by the emergence of a series of dark humps, centred at longitude 230°, on the north component of the SEB. On 27 February 1920 Harold Thomson, then director of the BAA Mars Section but also an avid Jupiter observer, wrote: 'The SEB is a most extraordinary spectacle. It consists largely of round dark dots and white spots.' The dark and white spots were moving retrograde, and 'the changes of aspect were so rapid that it was almost impossible to identify the markings after so short an interval as a couple of days.'[3] The entire region occupied by the faded SEB was now in a state of turmoil, turning into a maze of dark spots and White Ovals; dark material continued to spew from the source at 230° longitude and spread by wind shear from east to west at a rate of several degrees a day. During the following weeks this wave of darkening completely encircled the planet. At the end of this apparently eruptive process, the SEB had been restored to its original prominence.

Other violent outbreaks were observed in 1928 and 1943, at which point a new cycle set in, during which outbreaks occurred regularly at three-year intervals until 1958. There were further outbreaks in 1971 and 1975; the latter, which W. S. watched with keen interest with a 20-cm reflector, was particularly vigorous and involved an unprecedented four sources. There was another SEB disturbance in 1978, which was independently discovered by W. S., followed by a hiatus in the 1980s. Further outbreaks occurred in 1991 and 1993. The most recent, as of the time of writing, occurred at the end of 2010.

Noting that the SEB disturbances invariably begin with the sudden appearance of a small dark spot near the latitude of the middle of the South Equatorial Belt, Elmer J. Reese suggested in 1949 that perhaps they began as material reached the surface from an eruption of some kind. At that time, the rotation period of the planet's core, 9 hours, 55 minutes, 29.4 seconds – defining a third system of longitudes, System III – had not yet been discovered. (Radio astronomers first gleaned this from Jovian radio bursts in 1955.) Not until 1972 did Reese get around to re-analysing his data. He first tried to plot the source longitudes in System II. There was no clear pattern. He then replotted them in System III. This time he found the sources corresponded to three distinct loci (A, B, and C), slowly drifting with respect to the core and having a steady rotation period of 9 hours, 55 minutes, 30.1 seconds. When the 1975 revival occurred, it was found to be generally consistent with Reese's eruption hypothesis. Sources of the outbreak occurred within 2° of the longitude of loci A and B, with a third source, locus C, 23° ahead of locus B. (According to Rogers, the agreement was best if locus B had

Column 1:

Jupiter, 11 August 1928, as drawn by Antoniadi with the 83-cm Meudon refractor. The South Equatorial Belt is extremely faint, on the eve of its revival; the North Equatorial Belt is very broad.

Jupiter, 29 August 1928. Drawing by B. M. Peek, another leading British amateur observer of Jupiter, with a 32-cm reflector. This is an early stage of the great South Equatorial Belt revival.

A somewhat later stage of the South Equatorial Belt revival, as drawn by Phillips with his 50-cm reflector, 2 October 1928.

Column 2:

The South Equatorial Belt revival at its height, as drawn by T.E.R. Phillips, 5 November 1928.

T.E.R. Phillips drawing, 5 December 1928, shows the South Equatorial Belt revival continuing. The Great Red Spot, appearing as a White Oval, is on the extreme right.

Antoniadi drawing, 8 December 1928, showing the South Equatorial Belt revival at its height.

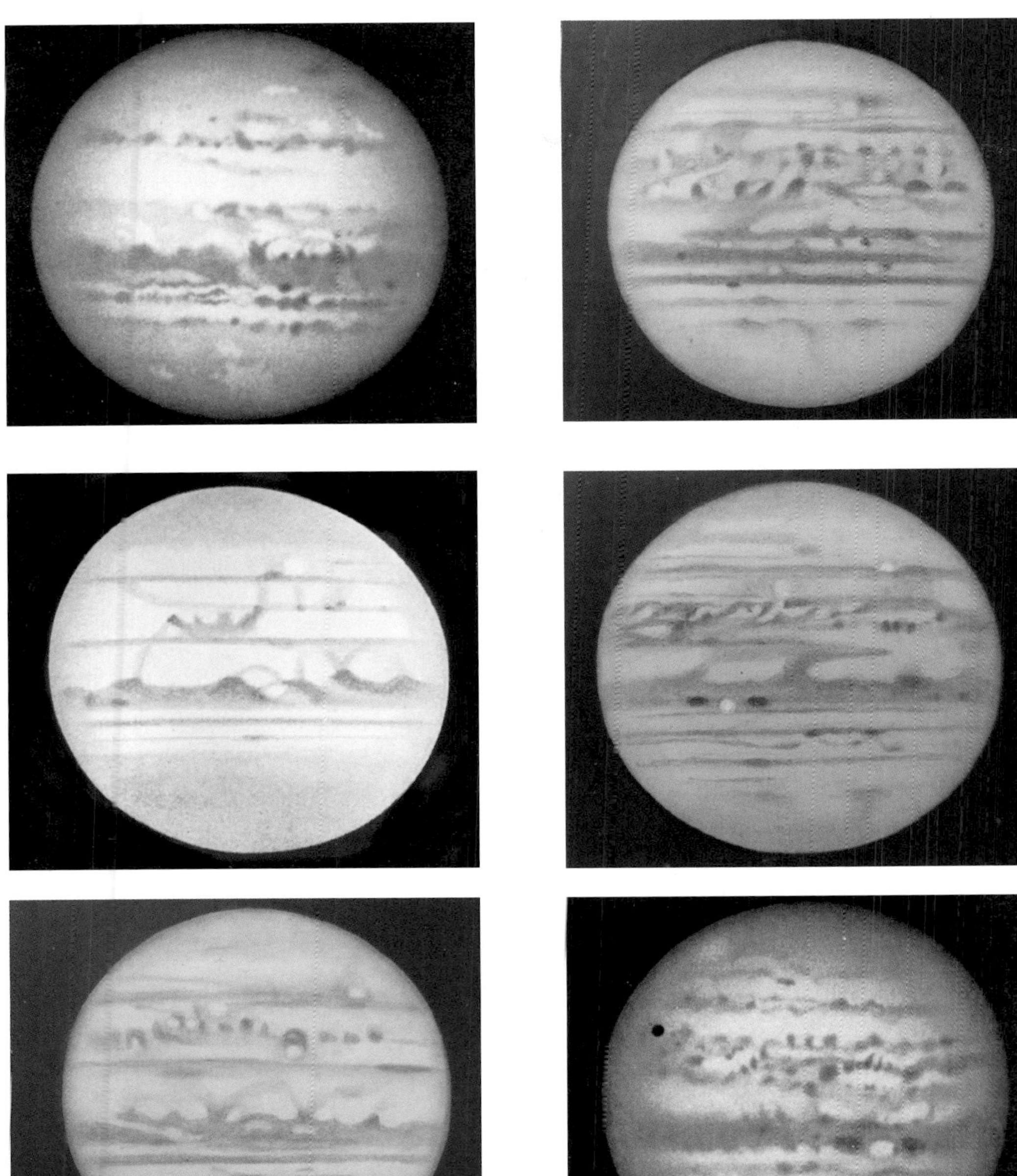

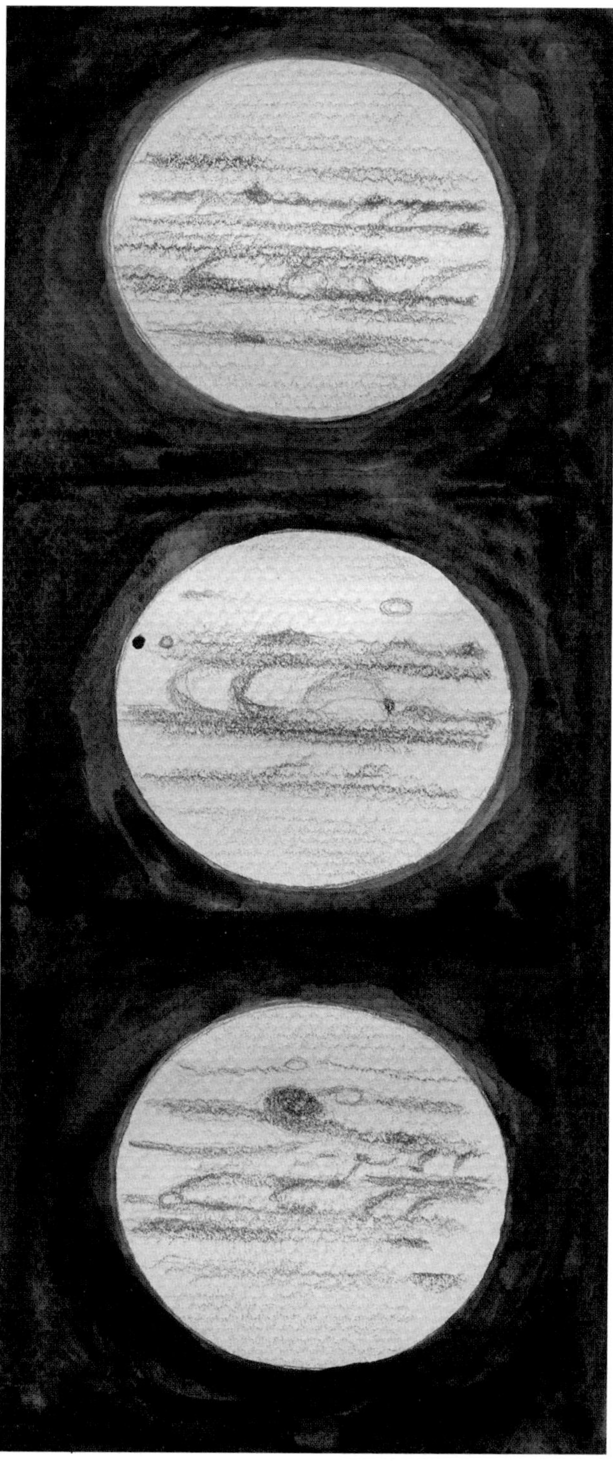

Stages in the development of a later South Equatorial Belt revival, that of 1975. From top to bottom: 27 August 1975, shows dark spots in the area of the source of the disturbance; 4 October and 10 October 1975, showing the revival in full swing. At this time there were also some prominent festoons extending from the fading North Equatorial Belt into the Equatorial Zone. Drawings by W. S. with a 15-cm reflector, and a magnifying power of 180x.

shifted 20° after 1958.) The sources of the 1990 and 1993 eruptions also coincided closely with the locus at B. It seems that locus A is the source of the most dramatic revivals (1919, 1928, 1971 and 1975), and locus C of the weakest. During revivals featuring multiple eruptions, as in 1975, the successive spot outbreaks occur sequentially – first from locus A, next from locus B and finally from locus C.[4]

We do not yet know the nature of the sources. Originally Reese suggested that they might be volcanoes, thus recalling Percival Lowell's idea of the Great Red Spot as a 'volcano in embryo' or of Richard Proctor's planet 'bubbling and seething with the intensity of primeval fires'. But BAA Jupiter Section Director John Rogers warns that they cannot be volcanoes in the usual sense. Instead, 'they might perhaps be long-lived circulations or waves or even floating objects at a deep level.' He adds that whatever Reese's sources may be, 'the instability always breaks first over them, just as clouds on Earth first form over mountains.'[5] The SEB disturbances – and the tortuous and convoluted cloud features associated with them – certainly provide us, observers on the Earth looking from the outside in, a rare glimpse of the violent processes that reign deep in Jupiter's interior.

In addition to the SEB revivals, the most impressive transient phenomena on Jupiter have included episodes of wholesale fading followed by revivals of the North Temperate Belt (NTB) and the North Equatorial Belt (NEB). The NTB revivals are heralded by the appearance of a very fast-moving plume on the NTB jet on the south edge of that belt. The NEB revivals, on the other hand, follow more extended disturbances. In recent years there have been NTB revivals in 2007 and 2012, which were also years when SEB revivals took place.

As of the time of writing – November 2016 – Jupiter has just emerged from superior conjunction with the Sun. As it becomes accessible to telescopic monitoring in the morning sky, the latest NTB revival is just getting underway, consisting of a series of bright spots on the belt. It will be interesting to see if, as expected, NEB and SEB disturbances will follow.

Though historically disturbances in the SEB and the NTB and NEB have been regarded as unrelated, according to work by Agustin Sanchez-Lavega, a planetary astronomer at the Higher Technical School of Engineering in Bilbao, Spain, and his colleagues, all have a similar convective origin.[6] This much seems clear. But a lingering mystery is why the SEB and NTB cycles often occur within a year of each other, and why they are often associated with intense colouring in the equatorial zone, thereby giving rise to an appearance that Rogers has called 'global upheaval'. (Note: the 'global upheaval' aspect is evident in the series of drawings by W. S. of the 1975 events.) Research continues.

The Modern View of the Vertical Structure of Jupiter's Atmosphere

We now know enough about Jovian conditions to be able to make sense of the painstaking record of phenomena made by passionate observers of the planet. We have already described the layers of clouds in the upper 60 km or so that we can study directly, from the chilly ammonia-ice cloud deck at the top down through what is likely the thick ammonium hydrosulphide layer of the brownish belts.

In zones between eastward- and westward-flowing currents, cloud masses expand and eddy in the wind. The eddies are highly unstable under the turbulent Jovian conditions, generally lasting only a day or two before being ripped apart by the violent zonal jets. But despite the turbulence of the cloud features, the zonal jets themselves appear to be remarkably stable – they have maintained nearly constant positions and wind speeds since Stanley Williams first mapped them in 1896.

The very deepest features visible in the Jovian atmosphere, at a depth of perhaps 100 km below the ammonia cirrus of the upper cloud deck, are dark projections – plumes or blue features. As noted

Fierce winds near the boundary between a Jovian belt and zone, as recorded by the Galileo spacecraft.

Vertical structure of the Equatorial Zone of Jupiter, based on Galileo data. The bluish areas are clouds at deeper levels.

above, for the last century these have been prominently displayed on the south edge of the NEB, though before 1911 they appeared on the north edge of the SEB. Their rotations are very close to System I's 9 hours, 50 minutes, 30 seconds, and they remain almost stationary at the System I longitudes they occupy. In all cases, they appear with their bases adjacent to the south edge of the NEB and look like dark spots or masses that in time develop graceful bluish festoons that project or loop into the EZ. As such, they are among the most characteristic and attractive features in this region of the planet. At any one time, there are usually twelve to fourteen of these plumes distributed at intervals of 25° to 35° in longitude. In the infrared, they show up as 'hot spots', which proves that they are actually cloud-free 'holes' extending through both the upper ammonia cirrus and ammonium hydrosulphide cloud decks. They extend to depths of about 100 km below the top of Jupiter's troposphere and through them heat escapes from below. Earlier students of the planet – including the late José Olivarez of ALPO, whose passion for these features led his colleagues to refer to them, unofficially, as the 'Olivarez Blue Features' – surmised that the bluish colour was due to these spaces being filled with upwelling crystals of water ice: in other words, ordinary snow coming from even further below.

A chance to test these theories directly occurred on 7 December 1995, when the Galileo atmospheric probe made its epic descent into the Jovian atmosphere, entering through a plume at the south edge of the NEB. This latitude, 6.5° North, corresponded to the border of the EZ and the NEB and marked the location of one of the most ripping jet streams on the planet. The probe was buffeted by even stronger winds at lower levels, thus proving – were further proof needed – that in contrast to the Earth, where cells of circulation are produced by solar heating, Jupiter's wind system is mostly driven by internal heat. The probe remained in contact with the orbiting mother ship for just under an hour; at the time contact was broken off, it had penetrated 155 km below the upper cloud deck, where the atmospheric pressure was 22 times that of the Earth at sea level (22 bars). Its signal lost, it disappeared beneath the swirling clouds forever.

During its windswept descent, the probe found little evidence of the threefold cloud layers – ammonia, ammonium hydrosulphide and water – that previously had been confidently hypothesized. Since the probe had entered into a clearing, this was not entirely surprising. However, the most significant finding – and this *was* unexpected – was the extreme dryness of the Jovian atmosphere. Previously it had been assumed that the planet might have an abundant endowment of water, perhaps tenfold greater than the Sun's; instead it appears to have less than the solar abundance. Might communication with the probe have broken off before it could reach the water?

That may well have been the case; however, if so, it has major implications for models of the origins of the planet. Instead of slowly accreting by sweeping up large numbers of small water-bearing objects such as comets, as formerly seemed plausible, Jupiter must have developed very quickly from the solar nebula. If, as the most recent calculations suggest, a Jupiter 'core' the size of the Moon formed within 100,000 years of the solar system's formation, and acquired close to its present mass of 318 Earth masses within a period of probably less than 10 million years, then from the first the planet was poised to take the lead role as the drama of the solar system unfolded.

Now that we have been able to characterize planetary systems round other stars, we can see that they show a great deal of variability. Our solar system, with a giant planet moving in a nearly circular orbit just beyond the water-ice condensation point (that is, at about 5 AU), turns out, contrary to expectation, to be far from typical. Naturally astronomers at first looked for systems analogous to ours and this caused them to miss what are actually more common arrangements. Some of the exoplanetary systems had widely spaced planets; others had several Uranus-sized planets spaced evenly together; but a common scenario, consisting of giant planets traveling in highly elliptical orbits very close to their stars (the so-called 'hot Jupiters'), was not even anticipated by theorists. Many of these systems are inherently unstable. Among the early exoplanet systems discovered, two – the Sun's system with four giant planets and the Upsilon Andromedae system with three – are stable. The worlds of our solar system have been orbiting safely for 4.6 billion years, while Upsilon Andromedae and its planets are estimated to be 2 or 3 billion years old, but even they seem to have barely avoided long-term instability. According to computer simulations by Rory Barnes and Thomas Quinn of the University of Washington in 2001, a surprising 15 to 20 per cent of planetary systems that start out fairly similar to these two will disrupt within a mere 1 million years. 'There is a suggestion,' they write,

> that, in general, planetary systems reside on this precipice of instability . . . It may mean that planetary systems tend to be as chock full of planets as they can possibly be – as if the planet-formation process works so well that an excess of planets is created, and some have to be flung away before the remainder can settle down into a barely stable pattern.[7]

Even in our solar system, it is clear that – as indicated earlier – the giant planets have migrated. Among the consequences of this migration was a fifth gas giant apparently being flung out of the solar system altogether. It is now careening somewhere through interstellar space, a vagabond world, isolated beyond our ability to imagine. As we just noted, right from the start Jupiter made its magisterial presence felt. Without that presence, as computer simulations of planet formation have shown, there would have been several massive planets ranging widely through the inner solar system – 'hot Jupiters' or at any rate 'hot giants' of the order of Uranus and Neptune, like those found in many other systems. As soon as Jupiter is included in the calculations, two terrestrial planets, along the lines of the Earth and Venus, tend to form, along with one small planet closer to the Sun (Mercury) and another at the orbit of Mars. But planet formation beyond the orbit of Mars becomes massively disrupted. There, instead of another planet, there formed a tattered band of rocky debris: the asteroid belt. Suffice it to say, the entire system of planets, including the Earth itself, bears the stamp of its most formidable member. The Earth's history – and perhaps its destiny – is intricately intertwined with that of Jupiter.

Below the Plumes

The pressure at the top of the yellow and tawny zones of the Jovian clouds is about equal to that of the Earth's atmosphere at sea level. At greater depths the pressures and temperatures rise steadily. There, the planet's internal structure is determined chiefly by the behavior of hydrogen under increasingly extreme conditions. Recall that, by volume, Jupiter consists of 90 per cent hydrogen, which is far and away the most common element in the universe, with helium a distant second. The Jovian ratio of hydrogen atoms to helium is, at 9:1, close to solar – and indeed cosmic – abundances. (Its atmosphere, by contrast, is a little short on helium.) Hydrogen behaves like a gas only in the outermost shell of the planet, to a depth of perhaps 1,000 km. At greater depths it liquefies; finally, at a depth of about 15,000 km, the pressure reaches 2 million times that of the Earth's atmosphere (2,000,000 bars), at which point hydrogen is transformed into a viscous metal – where by metal is meant a conductor of electrons. Only small quantities of this substance have been created in the laboratory and then only for a fleeting instant, because of the extreme conditions necessary to produce it.

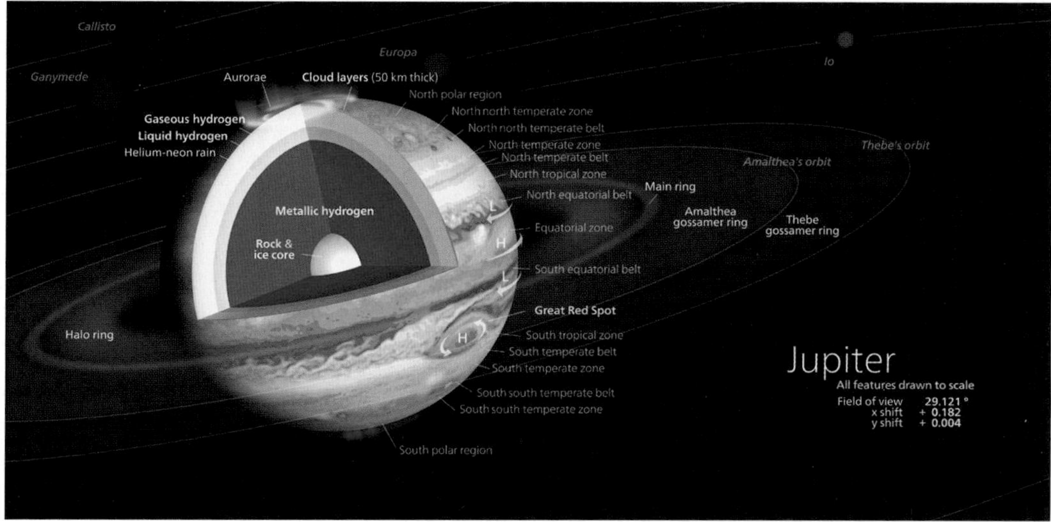

Jupiter in cross section.

The rotation of the 56,000-km-thick metallic hydrogen mantle produces powerful electric currents and generates an intense magnetic field. (Incidentally, the magnetic poles of Jupiter are 10° askew of the rotational poles; the reason is unknown.) The detection of faint radio noise by radio astronomers was identified with the core rotation. As noted earlier, this core rotation defines System III. A rocky and metallic core, of perhaps five to ten Earth masses, is thought to be harboured in the very centre of the planet and surrounded by a 3,000-km shell of hot 'ices'. This rocky and metallic centre is the seed about which Jupiter formed. The temperature here is 30,000°C and the pressure perhaps 100 million times that of the Earth's atmosphere at sea level (on the order of 100 megabars). The pressure here is so vast that the core would be crushed into a volume smaller than that of the Earth, so that even diamond – the least compressible substance known – would be squeezed to a density greater than that of lead at Earth-ambient temperatures.

Had Dante only known of Jupiter, he would have had a splendid place to situate his innermost circle of Hell!

SEVEN

ABOVE JUPITER

Though Jupiter's magnetic field is 20,000 times stronger than that of the Earth, owing to Jupiter's much greater size, the magnetic lines of flux are much more spread out so that at the top of the visible atmosphere – the closest thing to a surface a gas giant planet like Jupiter has – the magnetic force is attenuated to only fifteen times that of the Earth. (In addition, since Jupiter's magnetic field currently has the opposite polarity of the Earth's, a compass needle in the northern hemisphere of Jupiter would point south. Of course, magnetic polarity is not a fixed property; the Earth's last flipped about 780,000 years ago, while Jupiter's may have flipped as recently as 1911, when the polarity of plume features reversed from the north edge of the SEB to the south edge of the NEB, as described earlier.)

The planet's intense magnetic field largely shapes the nature of the environment around Jupiter and produces phenomena similar to the Van Allen radiation belts of the Earth. However, Jupiter's radiation belts are, like everything else Jovian, writ large. The intensity of radiation in Jupiter's radiation belts is so great that a hapless astronaut travelling through them would never live to tell the tale, as he or she would be exposed to 1,000 times the dose lethal to humans.

A gas of charged particles is known as a plasma. The plasma making up the solar wind is swept up and trapped in spirals in the magnetic field lines. As the lightest of these charged particles, electrons, are accelerated, they emit the violent bursts of electromagnetic radiation (synchrotron radiation) in the radio range that was first recognized by astronomers in the mid-1950s. Some of these electrons are accelerated to speeds approaching the speed of light. Since, as noted earlier, Jupiter's magnetic axis is tilted to its axis of rotation by some 10°, the radio signal 'wobbles'. It was the period of variation of this signal that was identified as the core's rotation.

A consequence of the solar wind's smacking up against Jupiter's magnetic field is the formation of a gigantic cavity referred to as the magnetosphere. Other planets – including the Earth – have magnetospheres; however, Jupiter's is much larger than any other planet's, and is the largest in the solar system apart from the Sun's. (Known as the heliosphere, it is the region of space through which the solar wind extends and the Sun's magnetic influence reaches. Its farthest limit, known as the heliopause, lies at the boundary where the solar wind encounters the interstellar medium, that is, the region of plasmas and magnetic fields of the rest of the Milky Way.) Jupiter's magnetosphere extends as much as 7 million km in the Sun's

The Jovian aurora on two different nights. The aurora is imaged in X-rays (Chandra satellite), the planet in visible light (Hubble Space Telescope). The aurora on Jupiter is much more intense than that on the Earth. Notice that it is occurring at both magnetic poles.

direction and almost as far as the orbit of Saturn in the opposite direction. It constitutes the largest continuous structure in the solar system apart from the heliosphere.

To put it another way, Jupiter's magnetosphere has a radius one hundred times that of the planet itself! Its volume is so enormous that, were our eyes sensitive to radio emissions, it would appear to subtend twice the apparent diameter of the Sun in our sky. It is easily large enough to encompass the orbits of all of the planet's satellites, and charged particles from their surfaces are swept into the magnetosphere. Jupiter's satellite Io, whose active volcanoes eject large amounts of sulphur dioxide gas into the space around Jupiter, forms a large flattened doughnut, or torus, of charged particles around Jupiter. The presence of this torus causes the radio bursts of Jupiter to be amplified by electrical discharges, essentially like lightning strikes that occur between Io and the planet. The intensity of the signal is thus enhanced whenever Io comes into certain positions in its orbit.

Charged particles trapped in Jupiter's magnetic field, however they come to be there, bounce between Jupiter's magnetic poles and, on interacting with the upper atmosphere, produce brilliant aurorae, counterparts of the Earth's Northern or Southern Lights. The difference is that, as might be expected, Jupiter's aurorae are far more intense. They have been observed across the entire electromagnetic spectrum, including the infrared, visible, ultraviolet and even the X-ray region; in the ultraviolet they are a thousand times brighter than the Earth's.

Jupiter's Rings

The planet Saturn is best known for its spectacular set of rings. At one time it was thought that it might be unique, in which case we on Earth would have been fortunate indeed to behold such splendour. As so often in the course of scientific advance, however, it turns out that ring systems are rather commonplace: in our own solar system not only Saturn but the rest of the giant planets – Jupiter, Uranus and Neptune – have rings.

After Saturn's rings, which were first recognized by telescopic observers in the seventeenth century, the next set to be discovered were those of Uranus, in 1977, followed by Jupiter's rings, revealed by the prolific Voyager missions in 1979–80. (There had been at least one premonition of their existence: as long ago as 1960, a Russian astronomer, S. K. Vsekhsvyatskij, suggested that a faint dark line seen right at the equator of Jupiter and referred to as the Equatorial Band might be the shadow cast onto the ball of the planet by an unseen Jovian ring. This was a not unreasonable guess at the time; however, it is now known that the Equatorial Band is a meteorological phenomenon and that the ring shadow is far too faint to be seen from the Earth.)

Indeed, Jupiter's rings are the definition of tenuity. They are exceedingly feeble, compared not only to Saturn's but to Uranus's as well, and only came to light as the Voyagers flew past the planet and looked back at it towards the Sun. From that hitherto unprecedented vantage point dust and ice particles showed up by forward-scattered light, in rather the same way that dirt and smears on a car windscreen are made visible as one drives into the sunrise or sunset. More recently, New Horizons, on its way to Pluto, managed to successfully image the rings for the first time ever from the sunward side.

The Jovian ring system has four components: a main ring, located between 123,000 and 128,940 km from the planet; a pair of very faint outer rings called the gossamer rings; and a torus-like inner ring called the halo. Each is a delicate, dusty structure and so exiguous as to be partially transparent. The fact that they forward-scatter more light than they reflect shows that the particles making them up must be tiny, like the particles making up a terrestrial fog.

Each of these rings has a sharp outer edge, but the inner edge is diffuse and extends all the way down to the upper atmosphere.

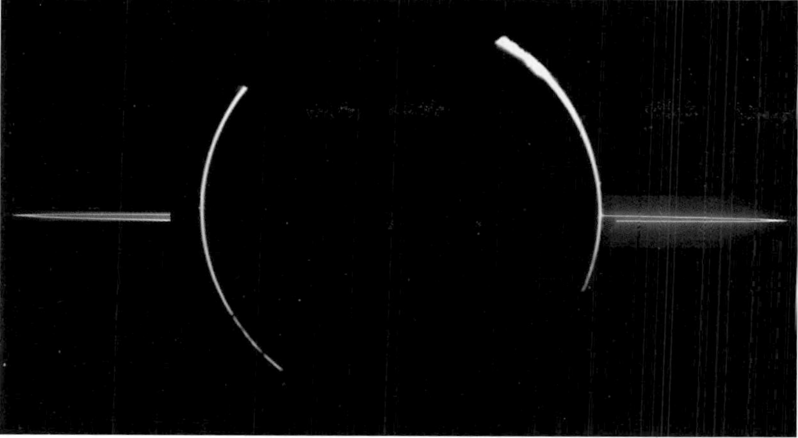

Galileo spacecraft image of backlit Jupiter's faint rings, taken on 9 November 1996

The tiny particles that make them up are thought to be dust thrown up by impacts on Jupiter's small satellites, whose orbits give definition to each ring's outer edge. Except for the halo, each ring has one or more associated satellites: the gossamer rings are bounded by the orbits of the satellites Amalthea and Thebe; Adrastea and Metis skirt the outer edges of the main ring. The tiny particles making up the rings cannot remain stable for long; the solar wind and the uppermost layers of Jupiter's outer atmosphere doubtless provide drag on them and send them spiralling inwards towards the planet below, where they are lost forever – burning up in a brief flurry of glory as meteors in the upper atmosphere and adding their pittance to the giant planet's mass. By such processes the rings would completely vanish within a few millions of years were they not replenished through the excavation of fresh material from satellite impacts.

How old, then, are the rings? Like ourselves, of which every particle of our composition is said to turn over every seven years, they are both young and old. Extremely young because the individual particles in the rings at any time must be relatively fresh additions due to impacts; extremely old since the small satellites that orbit close in to Jupiter and move in direct and nearly circular orbits, like Amalthea, Thebe, Adrastea and Metis, are believed to have formed from a circumplanetary disc that existed around Jupiter at the beginning of the solar system. Thus the Jovian rings, maintained in equilibrium through loss and regeneration, are likely to have survived since the beginning of the solar system, 4.6 billion years ago.

Jupiter's Satellites

The four largest satellites of Jupiter – all bigger than the Earth's Moon – were discovered by Galileo Galilei in January 1610, and named the 'Medicean Stars' after his patron, Cosimo de' Medici II. For a long time Galileo's claim to the discovery was disputed by Simon Mayr (also known as Marius), a German observer at Ansbach. Marius does seem to have discovered them independently and since he recorded the dates of his observations in the Julian calendar rather than in the Gregorian calendar used by Galileo, the dates of the notes of his observations appear to be ten days earlier than Galileo's. However, at that time dates in the Julian Calendar ran ten days earlier than those in the Gregorian, and once Marius's dates are converted it turns out that his notes began one day after Galileo's! A more important point is the fact that Galileo was far in advance in publishing his observations. His *Sidereus nuncius* appeared in early 1610, whereas Marius's *Mundus iovalis* did not appear until 1614. Thus – as bitterly noted by another rival of Galileo, the Austrian Jesuit Christoph Scheiner – by the time Marius's claim came out it was 'in vain and too late'.[1]

Marius deserves to be more than only a footnote in this history, since although the four satellites are known collectively as the 'Galilean' satellites, the individual names by which they are known were those given to them by Marius after the lovers of Jupiter. Their names, and their order of distance from the planet, can be remembered by the following verse by Marius himself:

As imagined by an artist, the scene in which Galileo demonstrates his telescope to the Doge and Senators of Venice. Note that Galileo's actual telescope looked nothing like that shown here. From Camille Flammarion, *Les Terres du ciel* (1884).

Relative sizes of the Galilean Satellites, New Horizons, February 2007.

Io, Europa, Ganimedes puer, atque Calisto
Iascivo nimium perplacuere Iovi.

Io, Europa, Ganymede the boy,
Callisto too did Jove with lust enjoy.

Parenthetically, Marius refrained from endorsing the Copernican theory, in which the Earth travelled around the Sun, on the grounds that the stars appeared to show discs in his small telescope. (The discs are now known to be spurious and are an optical effect.) If the stars showed discs they had to be implausibly large if Copernicus were right and the Earth travelled around the Sun. Instead Marius (and many others at the time) accepted the Tychonic theory, according to which the planets travelled around the Sun and the Sun in turn around the Earth.

Like Marius and other telescopic observers, Galileo also saw the small stellar discs. However, he did not take them literally. Instead he realized they must be due to some kind of optical aberration, though he did not grasp the precise explanation. As the best-known supporter of the Copernican theory of his day, he saw the Jovian satellites as an image, in miniature, of the solar system itself, and wrote in *Sidereus nuncius*:

> We have . . . an excellent and splendid argument for taking away the scruples of those who, while tolerating with equanimity the revolution of the planets round the Sun in the Copernican system, are so disturbed by the attendance of one Moon round the Earth . . . Here we have only one planet revolving round another . . . but our vision offers us four stars wandering round Jupiter like the Moon round the Earth while all together with Jupiter traverse a great circle round the Sun in the space of 12 years.[2]

In the small telescopes of the early seventeenth century, the Galilean satellites appeared starlike. Thus Galileo refers to them as 'stars', and calls them, collectively, the 'Medicean Stars'. The term

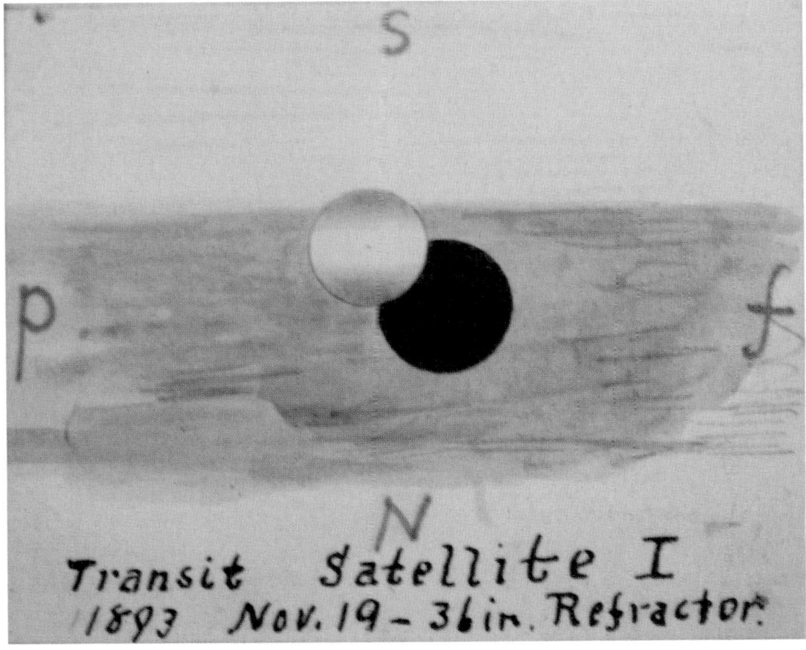

On 19 November 1893, E. E. Barnard captured the satellite Io, hovering above the shadow it cast on the Jovian clouds, with the 91-cm refractor of the Lick Observatory. Note that the equatorial region of Io appears brighter than the poles, which as we now know is actually the case.

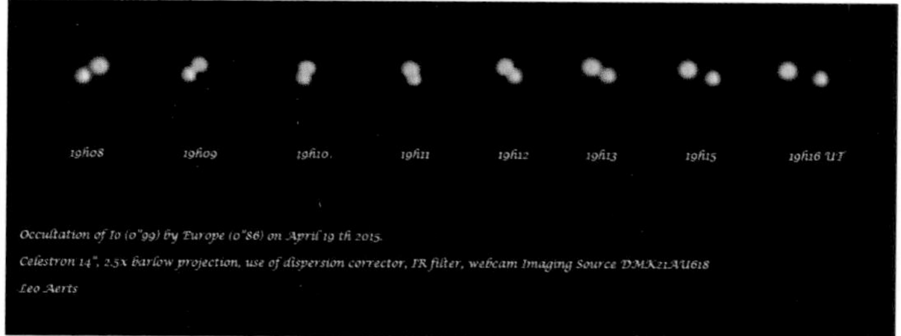

Europa occults Io, 10 April 2015. Images by Leo Aerts with a 36-cm Cassegrain and an Imaging Source DMK21AU618 webcam. Europa, which is the slightly smaller disc, appears sombre grey; Io, yellowish-gold.

Jupiter, 12 October 2013. Images by Leo Aerts, showing one of the infrequent triple satellite-shadow transits. Because three of the four satellites – Io, Europa and Ganymede – are in stable orbital resonances, with Io completing four orbits in the time Europa completes two and Ganymede one, two of these moons can appear in transit at once but never all three. For a triple satellite-shadow transit to occur, Callisto – the one moon not captured in an orbital resonance – must also be involved. As its orbit is slightly inclined to the equator, it usually misses Jupiter's disc from our line of sight, and thus the geometry permitting such events is realized only once or twice a decade.

'satellite' was coined by Johannes Kepler in a pamphlet, 'Narrato de observatis quatuor Jovis sattelitibus erronibus', published several months after *Sidereus nuncius*, in which Kepler described his own attempted observations. Despite having notoriously poor eyesight, he was successful. In fact, any modern opera glasses or binoculars will show them and it is even said that the two farthest out – Ganymede and Callisto – can just be glimpsed with the naked eye under unusual circumstances. (The authors hasten to add that they have never done so.) Even a 15-cm reflector begins to show the four Galileans as tiny discs, and it is perennially fascinating to watch the transits of the satellites and their shadows across the planet, the eclipses that occur when the satellites pass into Jupiter's shadow and the occultations when they pass behind the giant planet. (On rare occasions one satellite can even occult another.)

In small telescopes the satellite shadows appear inky black throughout, but each satellite has its own characteristic appearance in transit: Io and Europa are relatively bright and tend to be lost against the bright clouds, though they are easily visible in projection against duskier cloud features, while Ganymede and Callisto are so dark that at times they may be mistaken for shadows. 'Dark transits' (almost as dark as shadows) were often noted by the old observers, who thought there was a mystery here. However, the apparent darkness of the satellites as they traverse the disc is simply a matter of contrast with the brighter background clouds against which they are observed.

Historically the satellite phenomena have been of great importance. In the eighteenth century tables of the Galilean satellites' positions, calculated relative to a standard meridian

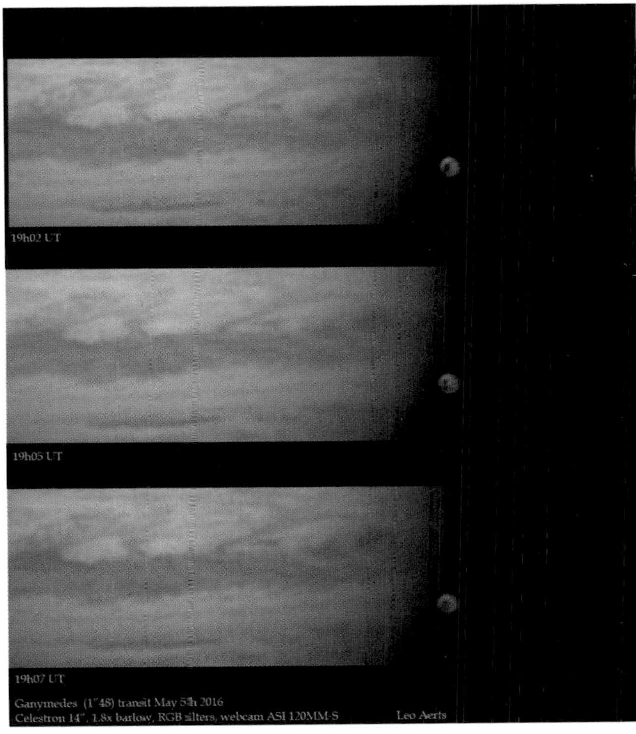

Ganymede about to transit Jupiter, 5 May 2016. These images, taken at 19:02 UT, 10:05 UT and 19:07 UT by Leo Aerts using a 36-cm Cassegrain, 1.8x barlow, RGB filters and a webcam ASI 120MM-S, show some of the characteristic blue features in the Equatorial Zone and dark markings on the surface of Ganymede.

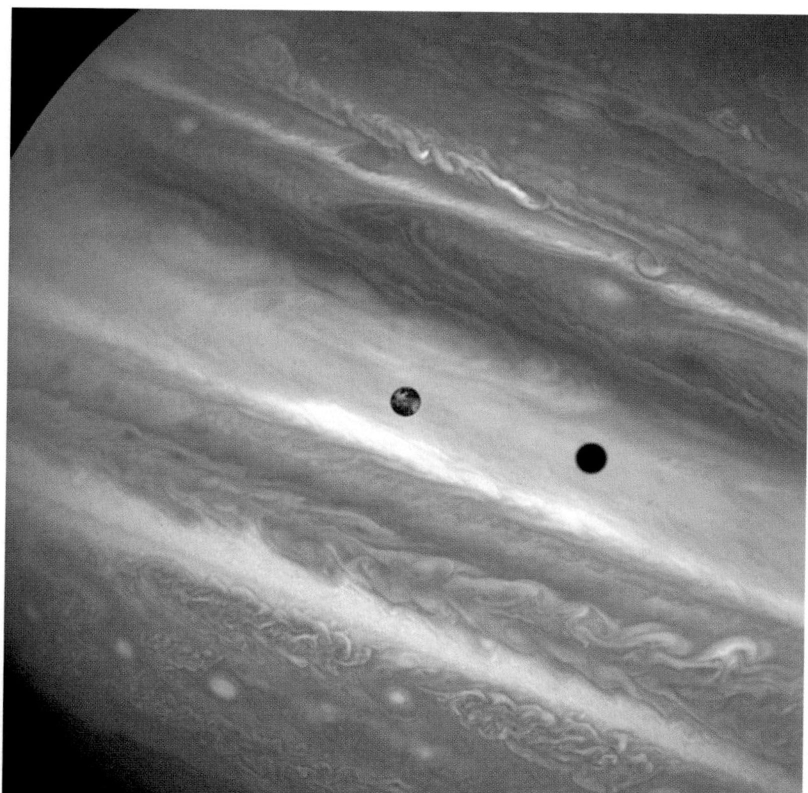

Io transits Jupiter. The black spot is the shadow of Io.

(for example, Greenwich or Paris), were compared with telescopic observations by navigators and explorers at distant points of the Earth's ocean as part of a scheme (one of many) to determine the observers' longitude.

In 1676, while timing satellite eclipses in order to improve these tables, Ole Rømer, a Danish astronomer working at the Paris Observatory, discovered a discrepancy between the predicted and observed times. In a stroke of genius, he explained this discrepancy by assuming that light – instead of travelling instantaneously, as had been thought previously – travelled with a finite velocity. It thus took a finite amount of time to traverse the distance from Jupiter, and this finite time would be greater as the planet's distance from the Earth was greater. As we now know, he was correct: light travels at 299,792,458 m/s (which defines our unit of measure, the metre) or 9.460730×10^{12} m/year (which defines the 'light year'). Because light can travel no faster than this – and neither can any other form of electromagnetic radiation, such as radio waves – even when Jupiter is closest to us, it requires a minimum of 32 minutes for its light to reach the Earth. When Jupiter is on the other side of the Sun, it takes almost seventeen minutes longer – or 49 minutes – to reach us. If an unseen object impacts Jupiter and creates a flash,

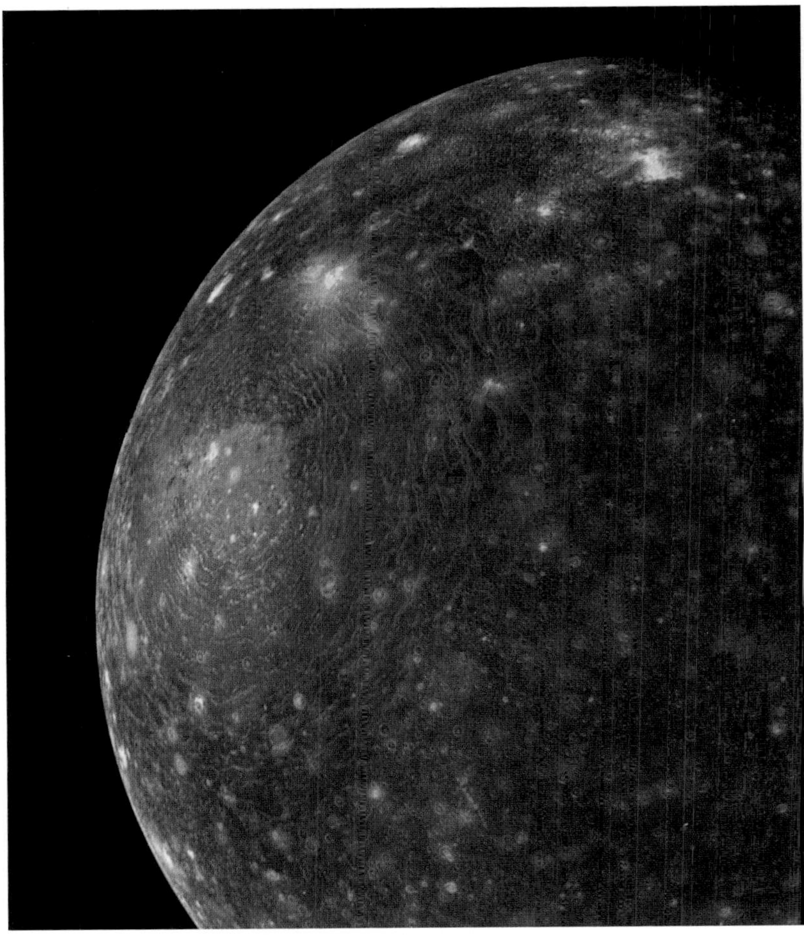

Callisto, as imaged by Voyager 1. The large brightish patch towards the left limb and just above centre is Valhalla, an ancient impact basin.

we will not actually learn about it until a minimum of 32 minutes after the event.

The Galilean satellites are large for their class and if they were not captured in orbits around Jupiter they would scale with the terrestrial planets. Thus Ganymede, the largest of the four (with a diameter of 5,268 km), is larger than Mercury (4,980 km), while Callisto (4,820 km) is just smaller. They appear small only compared to the giant planet they orbit.

The outermost of the Galileans is Callisto (1,883,000 km from Jupiter; orbital period = 16.7 days). Its density is about 1.8 g/cm^3. Ganymede is roughly the same, while Europa and Io are twice as dense at 3.0 and 3.5 g/cm^3 respectively. These satellites are clearly not gas giants, with densities around that of water, but neither are they terrestrial bodies. (The Earth's density, recall, is 5.5 g/cm^3.) From their intermediate densities and location it can be surmised that they represent another class of solar-system body, made up neither of rock and metal like the terrestrial planets, nor of gas like the giants, but of rock, metal and ice. Clearly, along this compositional spectrum, Ganymede and Callisto are relatively rich in ice – Callisto appears to be 50 per cent ice – while Europa and Io are rockier.

Ice is usually shiny and Callisto is not. If it is made up largely of ice, it is dirty ice, not the kind you would wish to find in your fizzy drink or wine cooler. Like the other Galileans, Callisto has differentiated into a water-ice (or, at warmer and insulated depths, liquid water) mantle surrounding a rocky/metallic core. Its surface is dark and peppered by craters – it boasts the highest crater density in the solar system. In fact, it is saturated with them, meaning a new crater cannot form except by obliterating or overlapping an existing one. Clearly, this can only be the case if Callisto's surface is very old – a palimpsest recording a series of explosive events caused by impacts involving smaller bodies scattered through the solar system since its beginning 4.6 billion years ago, and demonstrating very little capacity for turnover or resurfacing from internal geological processes. Callisto bears the scars of countless aeons of trauma. Apparently it has lacked the services of a good therapist.

Though at a glance Callisto resembles our own Moon, and has obviously been similarly heavily cratered by random chunks of rock and metal (meteoroids) and comets wandering the solar system, a closer look suggests that the craters look more like the divot remaining after ice is struck with a hammer than rock blown up with dynamite. Some of the features are on a grand scale and obviously record large-scale shocks. There are several multi-ring basins, like Mare Orientale on the Earth's Moon; the dominant one, known as Valhalla, is surrounded by broken ridges extending to a distance of 1,500 km from the centre. And yet even in the ridges of such features the relief is subdued. No doubt this is because – again – the crust is made up of water-ice, which has a lower tensile strength than rock. Despite the battered condition – and great age – of the surface, there is much less dramatic relief than in the case of Mercury and the Moon and the terminator appears remarkably smooth. As battered as its surface appears, many of the oldest features have no doubt been all but obliterated by flowing.

Each of the Galileans is unique. From Callisto's pockmarked shell of ice we turn, with relief, to Ganymede, the solar system's largest moon. Ganymede orbits at a distance of 1,070,000 km from Jupiter, with an orbital period of 7.15 days. Though Ganymede too has an icy surface, the structure is much more complex; there are primitive regions that are dark and heavily cratered and have been named for the discoverers of Jovian satellites. Thus there is a Galileo Regio, a Marius Regio, a Perrine Regio and a Nicholson Regio. The older, darker crust has in turn broken up into polygonal plates separated by regions of bright, grooved terrain known as sulci. The grooves lie parallel to the polygonal edges of the older terrain.

What happened here? There are several possibilities. One is suggested by simple analogy: as anyone who has filled an ice-cube

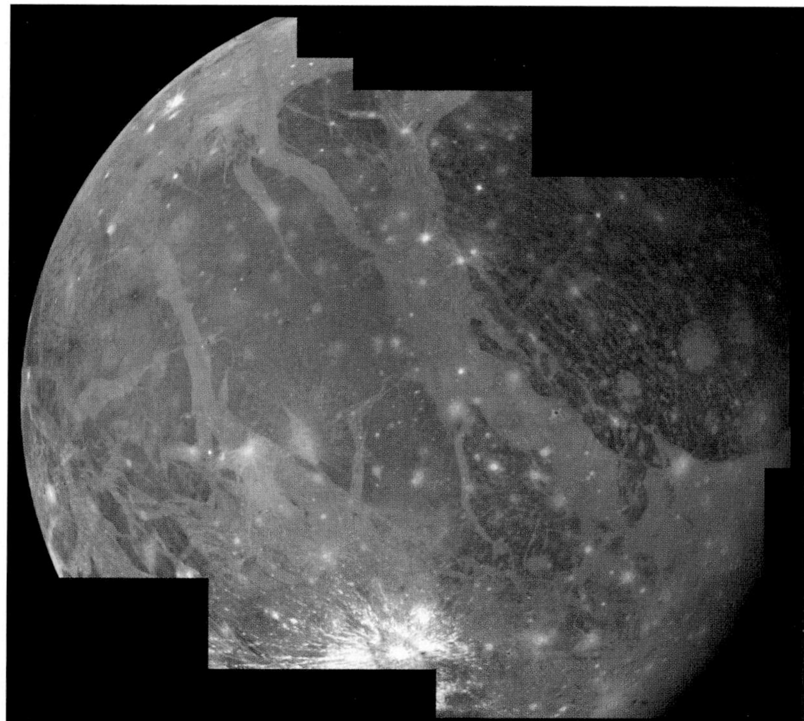

Voyager 2 mosaic of images of Ganymede. The hemisphere of Ganymede that faces away from the Sun displays a great variety of terrain. The large dark area to the upper right is known as Galileo Regio. The lighter grooved terrain below it forms bands of varying width, separating older surface units. The brightish area towards the bottom is the ray system of a crater that consists of water-ice splashed out in a relatively recent impact.

tray to the brim knows, ice expands as it freezes. Might not Ganymede, at some point in its history, have expanded? If so, the grooved terrain would have filled in where the old crust was insufficient to cover the satellite. In that case, the grooves are analogous to stretch marks. Another possibility is that the different terrains resemble the solid lava-filled basins and the highland regions on the Moon, but on Ganymede ice played the role of lava. Perhaps neither of these theories will prove to be correct, but of one thing we can be certain: though there is evidence of geological changes in the long ago past, geological activity on Ganymede has ceased. At least, in contrast to the relatively uninteresting terrain of Callisto – a history of sameness – that of Ganymede suggests a story.

With the Galilean satellites, the further in to the planet we approach, the more interesting the geology. The third, Europa

(671,000 km from Jupiter, with an orbital period of 3.55 days), wins the beauty pageant of the Jovian moons. It looks like a custom-made and carefully crafted blown-glass paper weight. Callisto and Ganymede are dark; Europa, on the other hand, has a relatively high albedo (0.64, which is almost as high as that of the notoriously brilliant, cloud-shrouded planet Venus). This in and of itself suggests that in contrast to its outward-lying siblings, Europa's surface is comparatively young – perhaps only 20 million to 180 million years old. Accordingly, there are but few craters on Europa; indeed, there is little relief of any kind, suggesting a relatively recent, clean crust. This crust is criss-crossed with cracks reminiscent of the thin ice covering parts of the Earth's Arctic Ocean. There are various explanations for this state of affairs. Perhaps Europa has not frozen all the way to the core. Perhaps there still exists a *liquid* mantle just beneath the visible surface of Europa. (This was made more probable by the 2013 discovery in Hubble Space Telescope images of 200-km-high geysers erupting near Europa's southern pole.) If so, Europa may be the only other solar-system member, besides the Earth, to have a global, briny ocean (though the other Galileans may have such crust-covered oceans, too, beneath varying thicknesses of ice). As such, Europa has become the most promising world beyond the Earth on which to look for life, and as such has attracted generous funding both from NASA and the European Space Agency, which are currently investigating the possibility of undertaking a series of Europa flyby missions to complete a detailed examination of its surface in the 2020s. (NASA's has been dubbed the Europa Clipper). One can imagine a future space probe that drills through Europa's ice layer (assuming that it is thin) and then submarines about, looking for what this ocean may harbour. We have only learned recently something about the nature of the Earth's ocean at its lowest depths. What lies hidden within Europa's?

After Europa, it might seem that anything would be an anticlimax. However, if anything, Io is even more captivating. It is arguably the

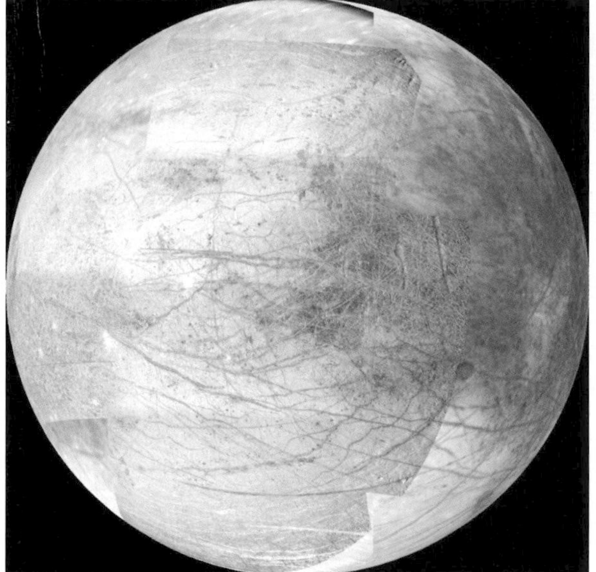

Europa imaged from the Galileo spacecraft.

Europa – the Jupiter-facing hemisphere. Mosaic of images from the Galileo spacecraft

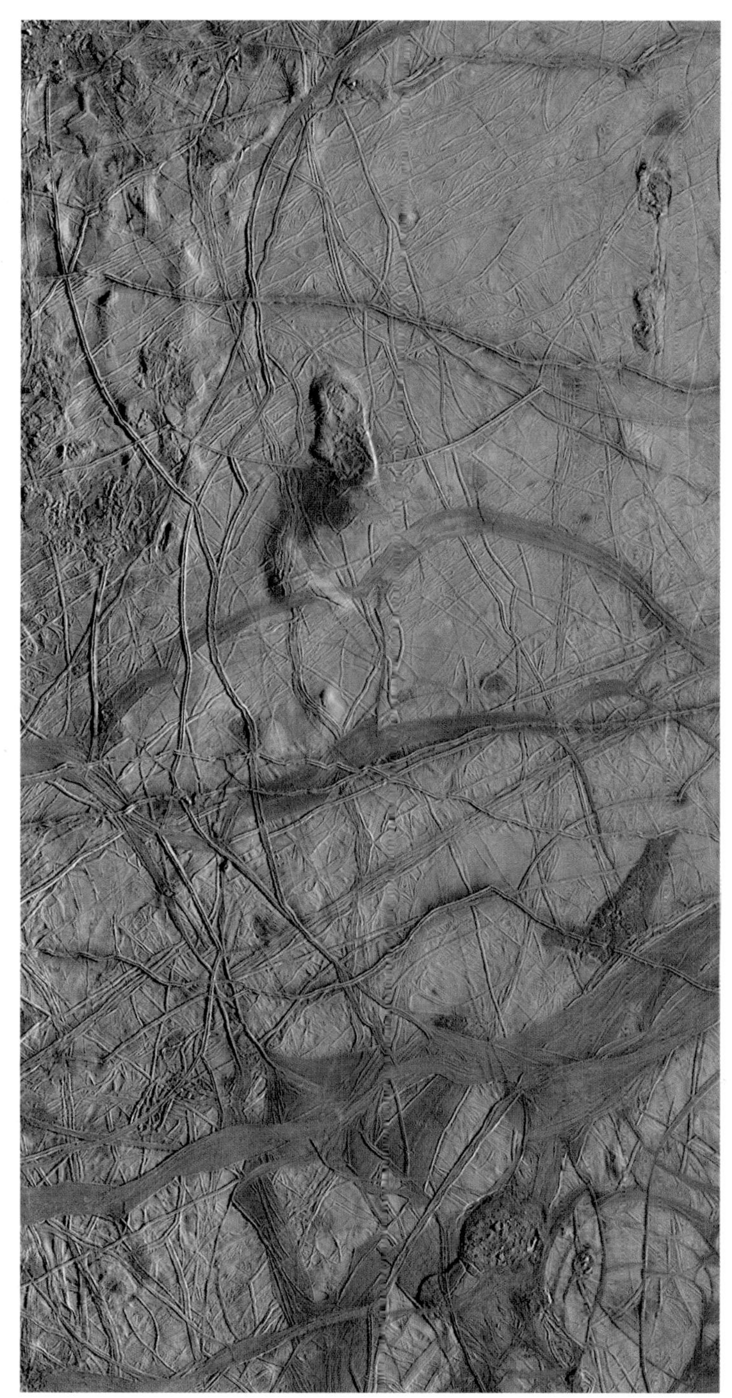

Colour-enhanced image of Europa's ice, from the Galileo spacecraft.

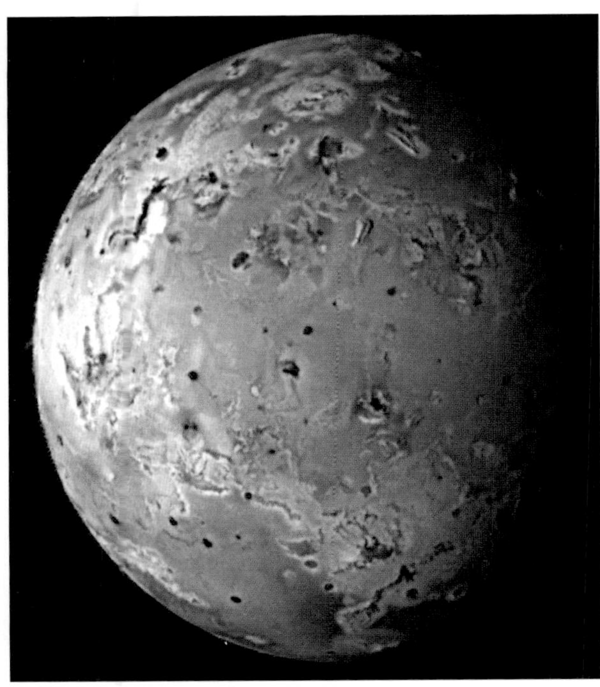

Colour-enhanced image of Europa's ice, from the Galileo spacecraft.

strangest satellite in the entire solar system. The innermost of the Galileans, Io orbits 422,000 km from Jupiter and has an orbital period of only 1.8 days. Thus on alternate nights it hops back and forth between the two sides of the planet. The other Galileans are almost colourless; not so Io. It is much more colourful than the other Galileans. This is apparent even in a smallish telescope. It looks decidedly yellowish. On close-up, the effect can no longer be denied, as it becomes almost garish; Io at close range looks rather like a pepperoni pizza, though one which has seen better days.

There are no impact craters on Io. Its surface undergoes constant renewal; thus it is as fresh as this morning's news, and there has simply been no time for impact craters to accumulate. There *are* round features that look like sink holes – and here hangs a tale. During the Voyager 1 flyby of Jupiter in March 1979, Linda Morabito, at the time serving as Cognizant Engineer on the Optical Navigation Image Processing System (ONIPS) of the Voyager Navigation Team, was examining the first photographs of Io sent back to the Earth, not to learn about its physical characteristics but to get a setting on its edge for navigational purposes. Thus she was looking at images in which the satellite was not centred, but the satellite's limb was well visualized. Much to her surprise, she found she was unable to fit a smooth curve to the satellite's limb; the reason was that the image contained an umbrella-shaped plume rising 270 km above the surface. It looked like an erupting volcano, and an erupting

Five views of Io, from the Galileo spacecraft.

volcano it was. It was a completely unexpected – and serendipitous – discovery. However, Morabito – at the time of writing an associate professor at Victor Valley College in Victorville, California – was not merely lucky. Io is constantly erupting – and continues to erupt even now. The volcanic 'sink holes' are all caldera.

Though coming as a complete surprise to most astronomers, the existence of active volcanoes on Io was not entirely unexpected. In a paper published just three days before the Voyager 1 flyby, Stanton Peale, Patrick Cassen and Ray Reynolds suggested that Io's interior might well be molten, owing to tidal interactions with Europa and Ganymede. In their own words:

> Calculations suggest that Io might currently be the most intensely heated terrestrial-type body in the solar system . . . One might speculate that widespread and recurrent volcanism would occur, leading to extensive differentiation and outgassing.[3]

Peale and his colleagues were building on the realization that, as had been known since William Herschel's time, all the Galilean satellites have captured rotations with respect to Jupiter, and travel in nearly circular orbits that lie close to the planet's equatorial plane. Moreover, the three inner moons – Io, Europa and Ganymede – are in a 4:2:1 orbital resonance. (Callisto, on the other hand, is odd-satellite-out.) In effect, whenever Io passes Europa or Ganymede, it gets a tug from them that pulls it slightly out of line. But Jupiter, never relinquishing its magisterial grip on its vassal, pulls it back again, like a dog on a leash. This planetary tug-of-war generates friction in Io's interior. The heat generated is sufficient to melt the rock. The lighter, more volatile elements have been forced to the surface long ago, and have boiled away; as a result, the outermost layers of its crust are rich in sulphur compounds, which account for the vibrant warm colours that splash across the satellite's surface.

A 'volcano' at the (top) limb of Io. New Horizons image, taken as the spacecraft passed the Jupiter system on the way to Pluto, February 2007.

At the low temperatures prevailing at Io's distance from the Sun, yellow sulphur ought to appear nearly white. When heated, however, yellow sulphur turns orange-red and becomes more viscous. At still higher temperatures it becomes a black liquid. All of these colours are found on Io.

During the Voyager 1 flyby, no fewer than eight active volcanoes were discovered. The largest – the one found by Linda Morabito – is Pele, named after the Hawaiian goddess of fire. The others were named Prometheus, Loki, Volund, Amirani, Maui, Marduk and Masubi. When Voyager 2 arrived four months later Pele had become quiescent, but the others were still erupting; indeed, Prometheus and Loki had become even more active. Ever since, the volcanoes have continued to be monitored from the Earth, and several more volcanoes have been discovered. Loki has been particularly active, and contains a black lake of liquid sulphur some 250 km across, within which float icebergs of solid sulphur. (Loki, officially known as Loki Patera, is the most powerful persistently active volcano in the solar system.)

We call these features volcanoes; in fact, however, they behave more like geysers, with each redistributing surface material from Io far from the source, given Io's low gravity (only about one-sixth of the Earth's, or comparable to that on the surface of the Moon). It has been estimated that gas and dust from Io's volcanoes are ejected at speeds of up to 0.5 to 1 km/s, which is much higher than in even the most powerful terrestrial volcanoes, such as Pinatubo and Krakatoa. Sulphur gas fuels the explosions from Pele; sulphur dioxide those from Prometheus and Loki.

It seems that new material may well up through faults on Io's surface. Moreover, the material spewed from its volcano-geysers has an interesting subsequent history. Much of it, of course, falls back to the surface, but some of it escapes and, colliding with ions trapped in Jupiter's magnetic field, becomes ionized. The ionized gas (plasma) forms a large ring surrounding Io, called the plasma torus.

Io is not only the most geologically active body among the Galilean satellites, but the most geologically active body in the solar system. It is far more active than the Earth. Though maps of Io have been produced by cartographers based on spacecraft images, they are little more than snapshot impressions, since by the time a cartographer has completed her work Io's surface has changed so much that a new map is called for. Maps from the Voyager era are as out of date as yesterday's pizza.

Io is splattered with streaks of black, yellow, red, blue and brown. These are the very colours of molecules containing the element sulphur. The key to everything on Io seems to be sulphur. In a chemistry lab, sulphur is a pale yellow. But that is at room temperature. Sulphur-containing organic molecules have colours depending upon the temperature at which they were 'cooked' inside Io.

All large satellites have tenuous atmospheres, but Io's is dominated by sulphur dioxide. Some of this sulphur-containing brew escapes the satellite: Amalthea, a small, nearby Jovian satellite, is orange. Apparently it is being dumped upon by Io.

Sulphur is an element we do not run into very often. Yet much of Io above an imagined rocky core is made out of it. It is perhaps incorrect to call Io an icy satellite. It is something else, a unique class of bodies in its own right.

Where does the energy come from to power such endogenic, internal behaviour? Our short list of energy sources has included proximity to the Sun and heat pent-up inside giant planets. Neither of these apply to Io. But notice that Io is the closest Galilean satellite to massive Jupiter. Its orbit is also the most eccentric of all the Galileans. The result is that the gravitational pull by Jupiter on Io is constantly changing, and not by a negligible amount. Meanwhile, Europa kicks Io with its gravitational foot, too. Sulphur does not have the hardness of rock. It is malleable. The changing gravity pushes and pulls at Io, stretching and squashing the satellite's interior sulphur like bread dough. Internal friction caused by all that sulphur rubbing against itself warms the satellite. Io's energy ultimately comes from Jupiter itself. Io is an energy thief.

(For an Io-energy analogy, try bending a paper clip back and forth until it breaks. Put the break to your lip. It is noticeably warmer.)

Did the Galilean satellites form as a miniaturized version of the solar system itself, with Jupiter playing the role of the Sun? Jupiter was certainly much hotter in the salad days of solar-system formation. Did it influence the composition of the Galileans? The densities of the Galilean satellites increase as we approach Jupiter. It may be that a once-hot Jupiter volatilized away some fraction of each satellite's ice, leaving the rocky/metal component behind. Io, then, would be the ultimate example of this, nearly stripped of ice. It is certainly hard to imagine that the sequence of satellites – which so resembles that of the solar system, from rocky terrestrial planets closer in to icier bodies farther out – is a coincidence.

In addition to the planetary-sized Galileans, Jupiter is attended by a huge retinue of lesser retainers. The fifth satellite, Amalthea, orbits inside Io. It was discovered by E. E. Barnard in 1892, almost three centuries after Galileo discovered the first four, and was the last satellite of the solar system to be discovered visually. (Incidentally, for a long time it was customary to refer to the Jovian satellites by Roman numerals, in order of their discovery; thus the Galileans,

In February 2007, as it flew past Jupiter seeking a gravity assist on its long journey to Pluto, the New Horizons spacecraft captured this montage of false-colour images of the giant planet and its satellite Io.

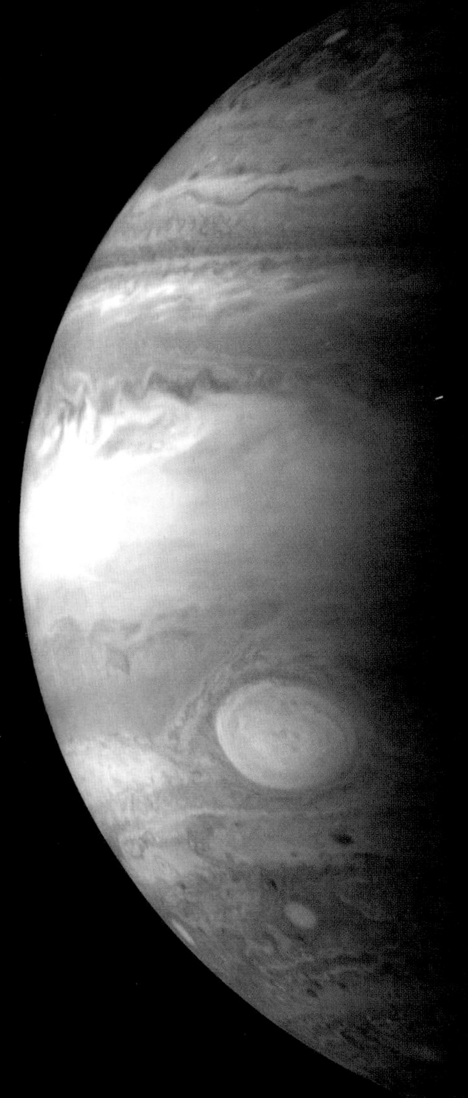

Io, Europa, Ganymede and Callisto, were referred to as I, II, III and IV, respectively; Amalthea was V, Himalia and Elara, discovered by C. D. Perrine of Lick Observatory in 1904–5, were VI and VII, and so on. Official International Astronomical Union-approved names were adopted only in 1973. Though 67 satellites are now claimed, most discovered on spacecraft images, only 53 are known well enough to have had their exact orbits calculated and to have received names. (It has been a chore to name them all and an even greater chore to remember them.)

In addition to Amalthea, there are three other inner satellites, discovered on Voyager images in 1980: Thebe, which orbits between Amalthea and Io, and Metis and Adrastea, both of which lie closer in to the planet and, as described above, confine the main Jovian ring, which is likely replenished by debris from ongoing impacts onto these satellites. All of these inner satellites presumably formed from remnants of the protoplanetary disc that surrounded Jupiter at the time of its formation.

Amalthea has been imaged by both the flying Voyager spacecraft and the orbiting Galileo. It is an oblong object measuring 250 × 146 × 128 km, of intensely reddish colour, redder even than Mars; this is presumably due to its being recipient of a coat of sulphur, courtesy of Io's ever-active volcanoes. It is of such low density that it is effectively little more than a porous pile of rubble and ice. It is heavily cratered; one particularly large crater, measuring 100 km wide and 8 km deep, has even received a name, Pan. (The name Pan is not entirely unproblematic, since it has also been given to a tiny satellite of Saturn.)

The only other Jovian satellite over 100 km across – and thus able to be glimpsed by amateurs equipped with large instruments – is Himalia (VI), which is one of four lying between 11,094,000 and 11,737,000 km from Jupiter's centre. (The others are Leda, discovered by Charles Kowal in 1974, Lysithea, discovered by Seth B. Nicholson in 1938, and Elara, discovered by Perrine in 1905).

In addition, there is a gallimaufry of outer satellites, all tiny and many of them irregular (that is, moving in retrograde orbits). They are asteroid-sized bodies, none more than a few tens of kilometres across, and were presumably captured by Jupiter over the long history of the solar system.

EIGHT

Jupiter in Collision

Acting like a giant vacuum cleaner, Jupiter's gravity has been singularly successful in gathering smaller bodies unto itself, as attested by the large retinue of small satellites just described. It also has a large family of short-period comets, by which we mean comets with orbital periods of less than two hundred years. These are thought to be icy objects captured from the Kuiper Belt beyond Neptune and as a class they are further subdivided into: 1) Jupiter-family comets, with periods of less than twenty years, whose orbits lie nearly in the plane of the ecliptic and do not extend much beyond Jupiter; and 2) Halley-type comets, with longer periods and more highly inclined orbits.

The best-known Jupiter-family comet is Encke, which has a period of only 3.3 years. However, there are many others; the latest count is about 520. They are listed at: https://physics.ucf.edu/~yfernandez/cometlist.html. Even this is doubtless an underestimate, since many more have presumably lost their volatiles during their repeated passages near the Sun. No longer able to outgas or form tails, they masquerade as near-Earth asteroids, recognizable for what they are – or were – only by the characteristics of their orbits.

The Kuiper Belt is a largely inexhaustible source of these icy proto-cometary objects. Indeed, it is estimated that in addition to the 100,000 or so Kuiper Belt Objects (KBOs) larger than 100 km that are believed to exist beyond the orbit of Neptune, there may well be a trillion or more of comet-nuclei size. For aeons, from that vast reservoir, icy bodies have been scattering inwards toward the Sun, crossing the orbits of the planets and meeting various fates, including colliding with Jupiter itself or with a member of its numerous retinue of satellites. (Because of its mass and the wide range of its gravitational grip, Jupiter and its satellites must have been assaulted especially often.) It is hardly surprising, therefore, that the icy surfaces of the outermost Galilean satellites, Callisto and Ganymede, boast the most heavily cratered surfaces in the solar system. It is certainly easy enough to grasp the dynamics of the situation – and the statistical odds of such events happening from time to time. Even so, one thing we personally never expected to see as fledgling amateur astronomers interested in the planets was an impact feature on Jupiter itself. After all, everyone knew that Jupiter has no solid surface! Then came the remarkable events of July 1994.

In March 1993, a new comet was discovered by the highly prolific team of Eugene Shoemaker, Carolyn Shoemaker and David Levy using the 46-cm Schmidt telescope at Mount Palomar. As their ninth discovery, it became known as Shoemaker-Levy 9 (SL9). It was soon identified as a member of the Jupiter family of comets. So far there was nothing unusual in any of this. What was unusual was that, at the time of its discovery, SL9 consisted of more than twenty fragments (none larger than a kilometre in size) voyaging through space in a line and popularly referred to as the 'string of pearls'. Working backwards, the comet, evidently little more than a loose pile of rubble, was found to have broken up the previous July as it brushed within only 100,000 km of Jupiter's centre. This meant that it had passed inside the orbit of Io and would have been subject to enormous tidal strains from the giant planet. Brian Marsden of the Harvard-Smithsonian Center for Astrophysics and director of the Central Bureau for Astronomical Telegrams, who was at the time the world's leading calculator of minor planet and comet orbits, soon followed up with an amazing telegram: the fragments were

This Hubble Space Telescope image shows the 'string of pearls', fragments of comet Shoemaker-Levy 9, before raining down upon Jupiter, 1994.

moving in a captured orbit round Jupiter and in fact were heading straight for the planet. This meant that the 'string of pearls' would proceed to impact, seriatim, the planet's southern hemisphere in July 1994. More detailed calculations showed that the SL9 fragments would come down on the planet's night side, at about latitude 45° South, with the first fragment 'A' due to hit on 16 July.

This was exciting indeed. As a regular observer of Jupiter since 1965, W. S. regarded the great South Equatorial Belt Disturbance of 1975 as the most remarkable phenomenon he had so far witnessed on the planet. Though the 1975 event was an outstanding example of its class, such disturbances had occurred before and would occur again. However, no one had ever seen a comet impact a planet or even the Moon. Jupiter's gravity would accelerate the comet fragments to a speed of 60 km/s (216,000 km/h), at which point they would explode in 10,000°C fireballs. Impact of the largest fragment would release the energy equivalent of 600 million megatons of TNT. (By comparison, the atomic bombs the U.S. dropped on Hiroshima and Nagasaki were in the mere 12 to 23 kiloton range, while the most powerful nuclear weapon ever tested, the Tsar Bomba of the former Soviet Union, was equal to only 50 megatons

The most exciting week in a lifetime for the Jupiter observer got underway on 17 July 1994, when the first SL9 fragments arrived at Jupiter. These drawings by W. S. that show the scene on 18 July at 5:22 UT and 5:36 UT were made using a 15-cm apochromatic refractor and a magnification of 300x. In the drawing at the top, the fragment 'A' impact site has rotated onto the disc, and in the next drawing has been joined by the fainter 'C' impact site.

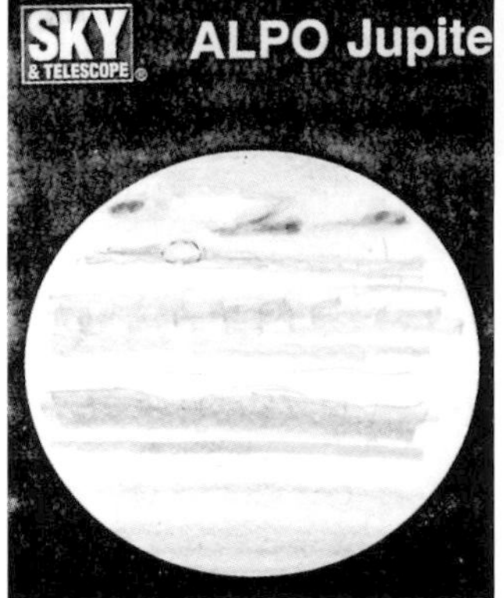

These drawings by W. S. with a 15-cm refractor, 300x, were made on 24 July 1994 at 3:30 UT and 4:00 UT, and show dark features of complex structure that have been produced by overlapping impacts. The feature in the centre of the lower drawing is 'UMWK', and was produced by the impacts of four fragments in nearly the same longitude of the planet. The feature coming onto the disc at the far right in this drawing forms the 'GDS' complex.

of TNT.) Clearly, each of the SL9 collisions would produce explosions thousands of times greater than those of all the nuclear weapons on the Earth combined, and it was expected that the ensuing shockwave would vaporize the impactor and some of Jupiter's atmosphere.

Remember that Jupiter never shows much of a phase defect as seen from the Earth. On 16 July Jupiter was 99 per cent illuminated; the collision sites occurred on Jupiter's night side, the side facing away from the Earth. However, they quickly rotated into view. (At the time, Jupiter was setting not long after the Sun; it was at a distance of 767,437,000 km or 42.7 light minutes away from the Earth. This meant that whatever happened, an observer on the Earth would have to wait 42.7 minutes after it was already over to find out about it.)

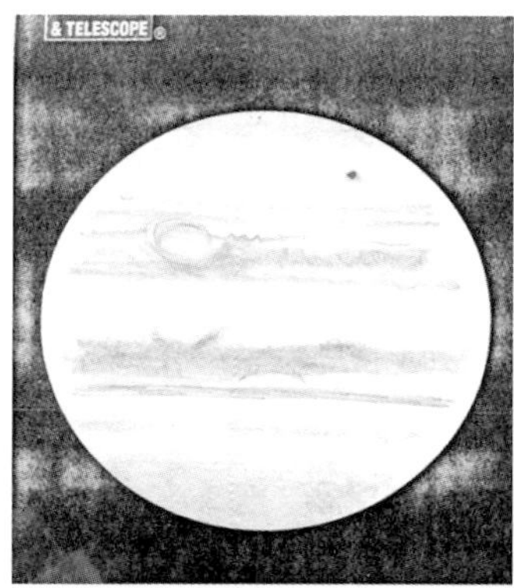

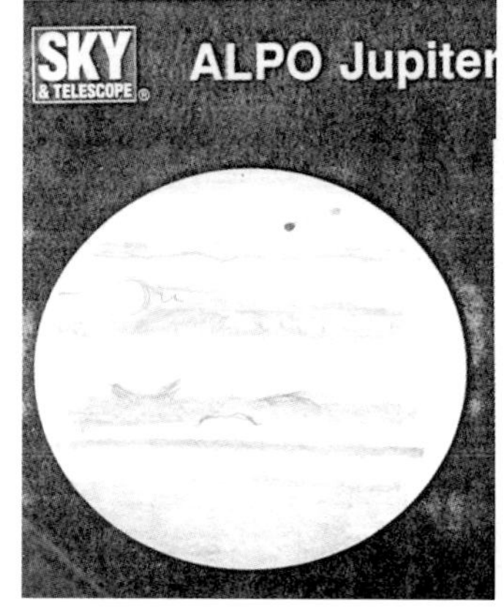

Further developments. The drawing on the top, made by W. S. on 20 July 1994 at 2:30 UT with a 15-cm apochromatic refractor, shows disruption of the SSTB by the expanding shock wave around impact site 'A'. The lower drawing, made on 22 July 1994 at 2:15 UT, shows accumulating impact features as more fragments impacted the planet. The curious bruise-like feature above and to the left of the Great Red Spot was produced by fragment 'H'.

W. S. began a vigil at the eyepiece of his 20-cm apochromatic refractor on the afternoon of 16 July, at just about the time fragment A was predicted to arrive. (The planet was easily found in the daylight sky because it happened to be quite close to the Moon at the time.) At once, he was startled to see an ink-black round spot, which at first he thought must be a satellite shadow. It was, however, larger and darker than any satellite shadow he had ever seen, and on closer inspection was found to be surrounded by a strange partial ring. The conclusion was inescapable: fragment A had arrived and made its mark! Though W. S. was unaware of it at the time, just three hours earlier the Hubble Space Telescope had recorded a plume of hot gas rising above the limb just after impact on the night side of the planet. (T. H. was observing Jupiter, too, with colleagues at the observatory of the University of Rochester – sort of a busman's holiday.)

The pearls of the string were coming down on average one every seven hours, and each one created a flash as it underwent instant incineration in the planet's stratosphere. Every night's observing session was awaited with eager anticipation. On 17 July, comet fragments C, D and E rained down, leaving, as fragment A had done, prominent bruise-like scars. On 18 July two of the largest fragments, G and H, arrived; fragment G did so near the earlier D impact site and produced a 3,000-km-high plume recorded by the Hubble Space Telescope, the resulting dark spot lying eccentrically within a pair of dark rings generated by a shockwave propagating

A fragment of comet Shoemaker-Levy 9 collides explosively with the Jovian night side. Galileo spacecraft images obtained en route to Jupiter, July 1994.

at 453 m/s (half the speed of sound in the Jovian atmosphere). Other large fragments, K and L, impacted the following day. For a few nights the prominent G and L spots were present on the disc at the same time and appeared, with their surrounding rings, like a pair of huge, black, ominously staring raccoon eyes. By the time the last fragment W struck on 22 July, the whole planet at latitude 45° south had been thoroughly carpet bombed, and was ringed with a line of dark complex bruises. Occasionally there were six or eight spots visible at a time, with the GDS and KW complexes being by far the most prominent – they could be made out easily in only a 5-cm refractor!

These spots, fully the size of the Earth, were by far the darkest ever seen on Jupiter. They consisted of fine debris – in effect, they were aerosolized comet corpses drifting back into the Jovian stratosphere from the point where the explosions had actually taken place, some 100 to 200 km above the visible cloud deck. Spectroscopic results from the Hubble Space Telescope showed that only the very largest fragments, such as G, had managed to penetrate the two upper cloud layers, and in doing so had dredged up ammonia, sulphur and hydrogen sulphide. Small amounts of magnesium, carbon monoxide and water were also detected, derived from the comet itself.

The discrete spots were quickly blown into tattered fragments by the east–west Jovian winds; inevitably they began to fade until by late September 1994 all that remained of the 'colossal wrecks' was a diffuse dark ribbon of material smeared through the entire impact zone, beneath which the Jovian cloud decks continued to roll on like a sea with heedless complacency. W. S. kept up his own observations until May 1995; by then the whole polar region above 45° latitude still appeared slate-grey or charcoal, presumably due to an admixture of 'soot' from the disintegrated comets.

All things considered, the effects of the SL9 crashes were rather minor and short-lived. And yet one could only imagine the devastation if the same fragments had impacted the Earth! Such an event would certainly have spelled the end of humankind and quite possibly even of life itself.

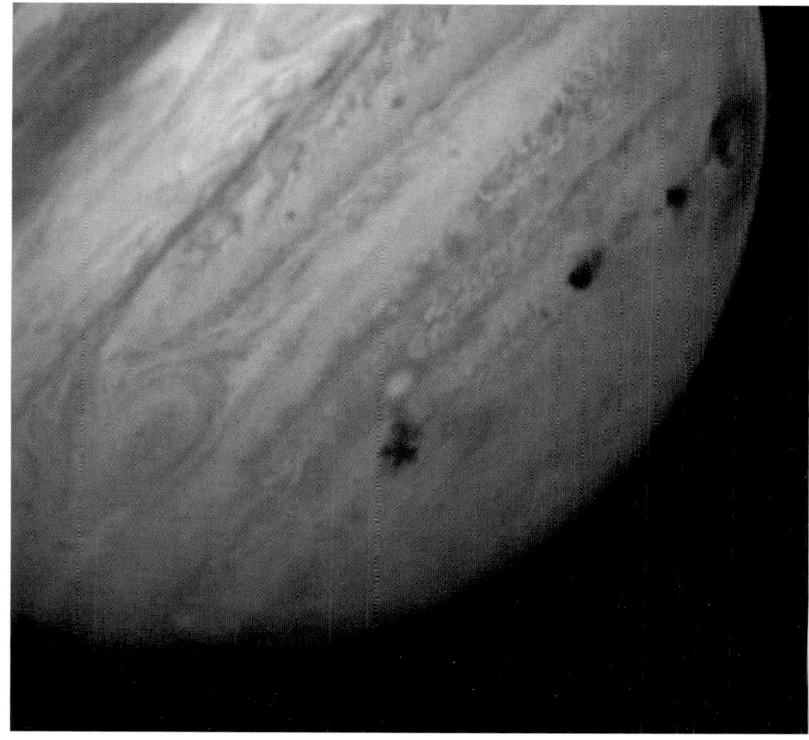

The series of bruises labelled, from left to right, GS, R, Q and H were easily visible in small telescopes, but the best view of events never to be repeated in a human lifetime were obtained by the Hubble Space Telescope, 23 July 1994.

Anyone who saw these momentous events felt privileged to witness a spectacle of cosmic significance – rather like viewing, from a safe distance, the KT-asteroid impact that may have wiped out the dinosaurs 65 million years ago. Just how often large objects like the SL9 fragments might hit Jupiter is uncertain. The best current estimates of the diameter of SL9 prior to its break-up indicate that it was only 1.5 km wide; the largest fragments would then have been about 500 to 700 m across. Eugene Shoemaker himself suggested that a body 1.5 km across might hit Jupiter once every hundred years or so, though collisions with comets that had already broken up like SL9 would probably be at least twenty times less frequent. A more quantitative analysis was attempted by David Kary (University of California at Santa Barbara) and Luke Dones (NASA-Ames Research Center), who used a computer simulation following 50,000 Jupiter family comets for 100,000 years. They 'observed' 750 simulated impacts (166 being from captured comets) and 1,052 Roche Zone disruptions, where the Roche Zone is the radius at which solid bodies would be ripped apart by Jupiter's tidal forces. They also noted thousands of captures. However, in their simulation there were only two cases where a captured comet was disrupted and then impacted on its next orbit like SL9 – the only mechanism, by the way, of generating a train of blows, since after another orbit the fragments would already have been too dispersed to do so. Thus their best estimates for mean intervals for 1-km objects (like fragment G) were impacts, 240 years; disruptions, 170 years; and SL9 'string of pearl'-type impacts, 90,000 years. Though these events would thus be rare in everyday terms – it is clear that we were exceedingly lucky to have witnessed them during our lifetime – they are hardly unusual over geological time scales, and similar comet-disruption events have left their mark on the Moon. The best-known example is the crater chain in the floor of Davy, a crater on the northeastern shore of Mare Nubium. No fewer than eight such chains have been found on the battered surface of Callisto.

In the aftermath of the SL9 impacts, the archives were ransacked in search of records of single-comet impacts buried among the thousands of observations of the planet. Needless to say, such events would be difficult to sift from records of ordinary dark cloud features.[1] As an aid in the quest, a set of useful criteria was proposed by co-author T. H.:

> Normally, Jovian cloud features show the strongest contrast near the central meridian. As they rotate toward the limb, the increased mass of overlying hazes in our line-of-sight reduces contrast, and the spots tend to disappear before they cross the limb or terminator (day/night line). This was not true for the SL9 spots. The material in the SL9 spots was deposited in the Jovian stratosphere. At the limb, an increased mass of spot material in the line-of-sight caused limb darkening. The only other Jovian feature that does not show relative limb-brightening is a satellite shadow, which displays constant contrast as it transits the planet.[2]

So far no feature in the archival record of the planet has satisfied these criteria. But then perhaps this is only to be expected. Until the second half of the nineteenth century, there were few people watching Jupiter at any time and their telescopes were, with a few notable exceptions, small. The twentieth century saw the heyday of conventional film (silver halide) photography, for which Jupiter – for reasons to be explained shortly – is a rather unsuitable object. Only with the advent of webcams and video imaging, which are the provenance of many highly skilful and diligent observers worldwide, was there any reasonable chance of an impact being recorded – and as we shall see, this has indeed been the case.

Further Jovian Impacts

Amateur observers of Jupiter have now recorded (as of the time of writing) at least four single-body impacts since SL9. The first was logged at round midnight on 19 July 2009 by Anthony Wesley, during an observing run with his 36-cm reflector at his observatory near Murrumbateman, Australia. Discouraged by the seeing conditions (atmospheric instability blurring a telescopic image) and about to pack up for the evening, but just before clicking 'exit' on his computer, Wesley changed his mind, and instead decided to wait for half an hour to see if conditions would improve. As soon as he returned to the telescope, he noticed a faint black spot in the South Polar Region near the eastern limb that had not been present when imaging the same region two days before. As soon as he convinced himself that he wasn't dreaming and that the feature was

A comet dies. Anthony Wesley, an amateur astronomer at Murrumbateman, Australia, obtained a video image registering the impact of an unknown object (probably a small comet) on Jupiter, 19 July 2009. The evolution of the 'bruise' – about the size of the Pacific Ocean on Earth when first discovered – was tracked by the Hubble Space Telescope from soon after its formation until it was utterly erased by the turbulent Jovian winds.

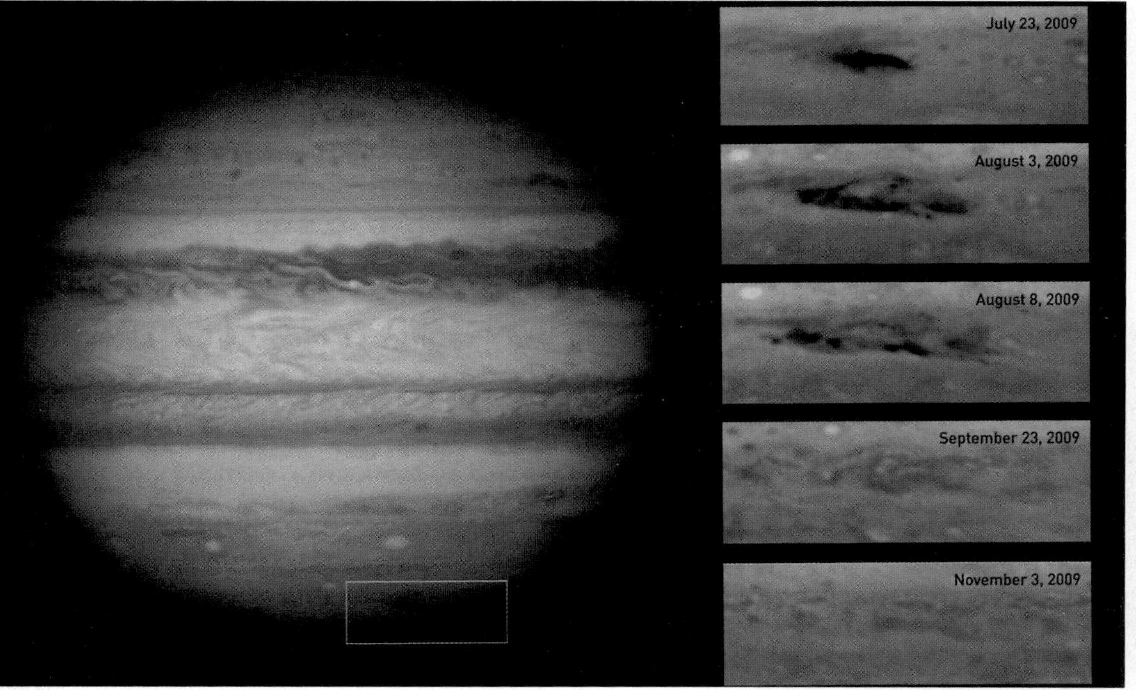

real, Wesley sent email alerts to other astronomers, including Glenn Orton, who twenty hours later used NASA's Infrared Telescope Facility on Mauna Kea, Hawaii, to confirm that it really was an impact. Using the Keck, Gemini North, Very Large Telescope and Hubble Space Telescope, astronomers monitored the evolution of the feature over succeeding weeks and, by working out the object's trajectory, established that it was more likely rocky than icy; in other words, it was an asteroid, not a comet like SL9.

Not quite a year later, while visiting a friend in Broken, Australia, Wesley was video imaging Jupiter and recorded another impact – this time as a flash – on 6 June 2010. It also was recorded by Christopher Go in the Philippines. The event was much smaller than the 2009 one and did not leave any mark.

Another flash event was recorded by Dan Petersen in Racine, Wisconsin, while he was visually observing Jupiter with a Meade 30-cm Schmidt-Cassegrain on 10 September 2012. This was the first visual sighting of a Jovian impact fireball. Petersen phoned Kyle Cudworth at the nearby Yerkes Observatory, and emailed Jupiter observers Richard Schmude of ALPO, John Rogers of the BAA and Christopher Go. He also posted a thread on the website Cloudy Nights, on which he announced: 'I observed an explosion on Jupiter this morning!' Among those who noticed Petersen's thread was George Hall, who had made a video recording of Jupiter that morning. On reviewing his video clip Hall found that he had serendipitously recorded Petersen's flash on 22 of his frames.

The fourth and latest fireball, lasting about one second, was detected on the limb of Jupiter on 17 March 2016 by video imagers Gerrit Kernbauer in Australia and John McKeon in Ireland.

Clearly, these events are more common than used to be thought and recording impacts on Jupiter is an area where amateurs still can make important discoveries. A concerted, worldwide effort to monitor Jupiter by mostly amateur astronomers suggests that up to 6.5 observable impact events may occur on Jupiter every year.

Those recorded so far have involved small asteroids or comets in the 10-metre range, comparable in energy to the smallest fragments of comet SL9 or to the Chelyabinsk meteoroid (Near Earth asteroid) that produced a dramatic fireball over the southern Urals in 2013. Such objects explode above Jupiter's cloud tops and – in contrast to the large SL9 fragments – leave no trace in the clouds themselves.

NINE

JUNO TO JUPITER

The latest spacecraft to explore Jupiter was launched from Cape Canaveral, Florida, on 5 August 2011. Juno – named after Jupiter's ever-jealous wife, who was able to peer through the clouds he enveloped himself with and thereby see his true appearance – actually travelled quite quickly to the distant planet, aided by a flyby of the Earth itself that borrowed some of our planet's gravitational potential energy. It entered Jovian space on 24 June 2016 as it passed through the 'bow shock', the distance from Jupiter at which Jupiter's magnetic field overpowers that of the Sun, and on 4 July 2016 became the second spacecraft, after Galileo, to slip into orbit round the giant planet.

The first Jupiter orbiter, Galileo, was launched from a Space Shuttle on 18 October 1989 and, after gravitational assist flybys of Venus and the Earth, arrived at Jupiter and was inserted into an equatorial orbit around the planet on 7 September 1995. It remained in orbit around Jupiter until 21 September 2003, when it was put on a collision course with Jupiter in order to avoid a chance collision and possible biological contamination of one of the Jovian satellites. In contrast to Galileo, which was powered by a radioactive source, Juno's energy source is entirely solar power. Also in contrast to Galileo, which remained in a near-equatorial orbit, Juno was sent into a polar orbit round Jupiter, in order to give scientists the first detailed images of Jupiter's poles. Following a 53-day elongated

One of the first images from the Juno space probe, 2016.

orbit around Jupiter, it made its first close approach to Jupiter (4,200 km from the cloud tops) on 27 August 2016. Originally mission planners intended to fire the major engine and drop Juno into a fourteen-day orbit, but when telemetry indicated possible problems with critical valves in the propulsion system, it was decided to play it safe – and keep Juno in its initial orbit. Though it will take longer to complete the mission objectives, scientists are confident that they will succeed in doing so. As of the time of writing (May 2017), Juno has just completed its sixth of 37 flybys. When the mission ends, in July 2018, Juno will be sent crashing into Jupiter – as with Galileo, a precaution to preclude any risk of biological contamination of the planet's satellites.

Though the mission is still in its early stages, it has already provided many important results.[1] It has provided the first detailed views of the planet's polar regions. (The only other spacecraft to provide non-oblique images of the planet's poles was Pioneer 11 in 1974, but at 10 times greater distance and with a much more primitive imaging system.) It was already well established from Earth-based observations that within 30° of latitude from each pole, where the zonal winds of lower latitudes drop, the familiar zonal banded structure breaks down and gives way to a dusky blue-grey 'polar region'. Juno shows it to be much bluer and stormier than hitherto imagined; the most distinctive features are bright ovals with spiral extensions, which time-lapse image sequences show to be cyclones – rotating counterclockwise in the northern hemisphere and clockwise in the southern hemisphere. A number of these cyclones cluster round each pole.

Other bright regions are more nondescript, and resemble the small chaotic features familiar from Voyager images of low-latitude clouds. A number of the White Ovals cluster round the poles in each hemisphere – though the patterns at each pole are markedly different; as yet it is uncertain whether these features exist in stable equilibrium, or whether they are dynamic and will evolve over time. Another important Juno result is that the circumpolar waves on Jupiter do not give rise to a north polar hexagon like that on its neighbour Saturn, proving that the polar dynamics and atmospheric structure of these planets are distinct. Another dramatic result was a high-altitude cloud, or possibly a detached haze layer, which cast a shadow on the main polar cloud deck some 60 km below. (The only other planet where this has been seen was Neptune, where during its August 1989 flyby Voyager 2 recorded shadows cast by high-altitude cirrus clouds of methane ice onto the bluish methane clouds below.)

In addition to images obtained by the visible-light JunoCam instrument (with blue, green, red and methane filters), thermal emission data providing information about unexplored regions of Jupiter's deep atmosphere was captured by the spacecraft's microwave radiometer (MWR) and Jovian Infrared Auroal Mapper (JIRAM). Before Juno, almost all that was known about Jupiter's atmosphere was based on observations of the upper cloud tops (that is, a region where the pressures were 0.5 bar or less; there was, as noted earlier, one series of measurements to the 22-bar level at a single point on the planet provided by the Galileo probe, but that was all). Though it was often assumed that below the cloud tops, in the sunless depths, Jupiter would look much the same wherever it was probed, there was really no evidence one way or the other. Though still early in its mission, Juno has already shown that in fact this is far from the case, and that below the cloud tops things are actually very complex. The dominant feature is the equatorial plume and the neighboring NEB which, at least down to the 60-bar pressure level, resemble a system similar to the Hadley cell on Earth, in which air rises near the equator, flows polewards, sinks again at roughly 30° latitude north or south, and then returns equatorward near the surface. (On the Earth, the Hadley cell is responsible for the tropical trade winds.) Something similar has long been surmised to explain the low-latitude circulation on Jupiter, but again, things are different there. On Jupiter, ammonia plays the role that water does on Earth; ammonia rising in the equatorial plume forms crystals but the crystals falling out evaporate at the 1.5-bar pressure level, so on Jupiter the system is a Hadley cell without rain.

The JIRAM instrument also provided information about areas where thermal radiation escapes from pressure depths greater than a few bars. The 'hot spots', which represent regions of downwelling and dry air (where the relative humidity is lower than 3 per cent) include, most notably, the zone between 5° and 20° within the NEB, as well as the other main belts. The latitudes, evidently, are cooler, moister regions, characterized by upwelling. The Great Red Spot and White Ovals are also areas of upwelling.

It is clear from Juno's results that rather than being a mere skim on the surface, Jupiter's belts and zones have deep roots extending downward hundreds of kilometres. Another important result, from Juno's magnetometer investigation (MAG), is that the planet's magnetic field, already known to be the most intense planetary magnetic field in the solar system, was found to be almost twice as strong as expected. The electrically conducting fluid in convective motion that forms the dynamo responsible for generating the magnetic field must, therefore, be located in a region not far below the surface – an unexpected result. At present, the best guess is that it occurs in a molecular hydrogen layer above the region where increasing pressure in the Jovian depths induces transition to metallic hydrogen. Finally, from its polar orbit, Juno has made many observations of the Jovian aurorae, where different processes seem to be involved than with the aurorae on Earth.

These are all fascinating details, and Juno's initial results have already changed our ideas about Jovian conditions and begun to answer some of the monumental questions. By the time the mission ends in July 2018, it is hoped that we will have (at least tentative) answers to the following:

1. How much water is there really in Jupiter? Knowing this will help decide among competing theories of Jupiter's formation. Did hydrogen-helium Jupiter form all at once? Or was it collected by an already existing core? What was the role of icy planetesimal impacts, objects which might be expected to deliver water to Jupiter – and to the early Earth?
2. What are the composition, temperature and cloud motions of Jupiter's deep atmosphere? Unlike the Earth, Jupiter is still in a primordial state. What do the depths of Jupiter today tell us about our own planet's earliest days?
3. What is the detailed structure of Jupiter's magnetic field? Of its gravitational field? Mapping these structures will in turn provide us with better models of the Jovian interior, including, perhaps, the answer to the question of whether the core is solid or liquid (or consists of layers of both). We would also like to know the core's size, and of what it is made.
4. How does Jupiter's magnetosphere behave? How does it entrap and channel into the upper atmosphere the charged particles that give rise to the polar aurorae, which are the most spectacular in the solar system?

Juno's highly elliptical, polar orbit means that it will not be well placed to study the equator-loving Jovian moons. For that exploration other missions are planned, including the European

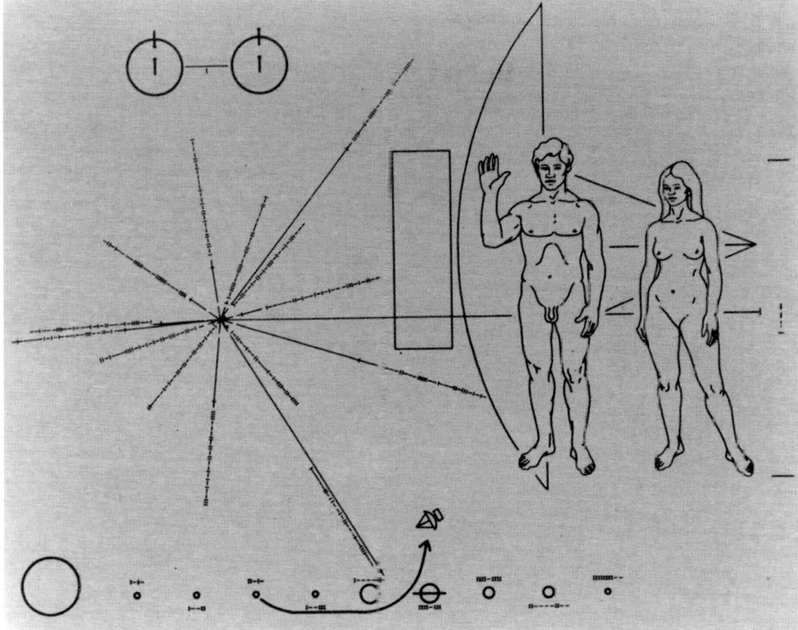

The Pioneer plaque. The 'return address' is indicated by lines radiating from us to a unique kind of star called a pulsar.

Space Agency's Jupiter Icy Moons Explorer (JUICE), scheduled to be launched in 2022 for a 2030 arrival in orbit around Ganymede. JUICE's mission will be to add to our understanding of the conditions for planet formation and perhaps of the origin of life. Meanwhile, NASA has prioritized its own mission to Europa, the satellite deemed most likely to harbour life, with a tentative launch date in the late 2020s.

These are exciting plans, but we must not forget the missions that have gone before. Cassini and Juno have, of course, achieved what not long ago would have seemed unthinkable, and entered orbit around the most Olympian of the planets. And before them were the Pioneer and Voyager flyby missions. They encountered Jupiter while outward-bound from the solar system altogether – the Pioneers equipped with a plaque designed by Carl Sagan and his colleagues to convey useful information about the third planet from the Sun (the Earth) and the beings that made it (ourselves),

This view, created by amateur scientist Roman Tkachenko using data from the JunoCam imager on the Juno spacecraft, looks directly at the south pole of Jupiter, 2 February 2017. The spacecraft was then 102,100 km above the cloud tops. Cyclones swirl around the south pole, and one of the White Ovals can be seen at the extreme limb at lower left.

the Voyagers with a phonograph record containing sounds and songs of the Earth. Of course, space is so vast that it is unlikely that any intelligent alien race will ever encounter these artefacts – just as the chances are almost nil that we would ever notice any sent our way from an extrasolar planet.

The last signal from Pioneer 10 was received on 22 January 2003, from far beyond Neptune, but the Voyagers live on: a significant landmark occurred on 25 August 2012, when Voyager 1 officially passed the heliopause and entered the realm of interstellar space.

As our flyby emissaries continue their journey, they no longer roam among the planets of our system but wander into the realm of the other stars. Many, as we did not know when they set out but now do, have planets of their own. Already, more than two thousand are known, and their number must ultimately be unfathomable. Many of the earliest ones – which were completely unanticipated by astronomers – are high-mass planets that *closely* approach their parent stars (less than 0.5 AU, where 1 AU is the distance from the Earth to the Sun). These are universally referred to as 'hot Jupiters'. These planets have atmospheric temperatures of thousands of degrees Celsius, so that clearly they could not have formed in their present locations. Evidently they have migrated round their systems – as Jupiter and the other giant planets are believed to have done in the early history of the solar system.

Juno image, 11 December 2016, during the spacecraft's third flyby. The spacecraft was 458,800 km from Jupiter at the time. This view, taken over the southern hemisphere, shows the Great Red Spot near the upper part of the planet's crescent, and just below that the Red Spot, Junior. Near the middle of the disc the string of White Ovals marches from upper left towards lower right in the image.

TEN

Observing Jupiter

Even if one looks at them through binoculars, Jupiter and its four large satellites are rewarding objects. The moons are ever-changing with their eternal dance, and one cannot witness the scene without remembering Galileo's triumphs and all his woes. With Jupiter's large apparent diameter, the extended periods during which it can be profitably observed and the endless fascination of its various and ever-changing phenomena, the planet itself remains – even in the Galileo and Juno era – the 'amateur's planet'. The interested and motivated individual can still contribute to the venerable observing programs of the Jupiter sections of the BAA and the ALPO and, particularly if equipped with video- and CCD-imaging capabilities, can glean data of lasting scientific value.

It is worth remembering that Stanley Williams and Elmer J. Reese, whose names stand very high in the list of students of the planet, did most of the work for which they are remembered with reflectors of less than 20-cm aperture. One of the authors (W. S.), while at home from college in March 1978, made an independent discovery of a new SEB disturbance with a 20-cm reflector. There was nothing extraordinary in the feat; it was simply a matter of looking at the right time and knowing enough to recognize the significance of what was seen.

In starting out, the student who would like to make a serious study will find it helpful to devote some time to learning the nomenclature of the belts and zones, including their abbreviations. Obviously the features shown in the diagram represent the mean appearance, and since Jupiter is a planet of cloudforms – often undergoing rapid and dramatic changes – an actual observation will show significant departures from the mean.

The belts and zones are visible even through small telescopes. Larger apertures begin to show that they are far from uniform. The edges of the belts appear decidedly ragged, with the interiors often hollowed out. Some features are best referred to simply as 'spots', though in some cases more descriptive terms, first introduced into the literature of the planet by the classical observers, may be useful: dark and light features on the edges of the belts, for example, are frequently referred to as dark 'projections' and light 'bays'. The projections may be stubby and short or may extend into long streaks or 'wisps'. If noticeably curved, the streaks are known as 'festoons'. A festoon with a bright white spot at the base is called a 'plume'. Short, very dark streaks, often brownish in colour, are called 'barges', large elongated spots are called 'ovals' and so on. (Though all of these terms provide convenient shorthand descriptions, they are at most semi-official. Nor do they – especially in the case of 'barges' – imply any surmise as to the nature of these features!)

Though much of the amateur monitoring of Jupiter is now done with CCDs, avid visual observers of the planet still exist, and still do useful work. Most observers will want to attempt a few whole-disc sketches in order to record the overall appearance for a given apparition. A prepared disc giving the dimensions of the planet's oblate outline will be convenient, such as the one provided overleaf that is suitable for copying. If recent measures of the latitudes of the belts are in hand, it may be convenient to sketch the positions of the belts lightly before venturing to the telescope. Unless there is a compelling reason to do otherwise, it is best to study the planet for half an hour or so before beginning a sketch, in order to get a sense of the 'seeing', that is, the quality of atmospheric conditions. Unless the air is steady and the definition good, there is not much point to proceeding further. This will also allow the positions and forms of the most conspicuous markings to be laid down. Once a sketch is begun, it is necessary, of course, owing to the rapidity of Jupiter's rotation, to put in the main features quickly, but once these have been established, the finer features can be entered with reference to the main features, and at greater leisure.

In general, it is probably more useful to produce detailed sketches of interesting parts of the planet or strip-maps showing in detail the features that are brought into view over several hours at the eyepiece rather than full-disc drawings. Among the most conspicuous and interesting features in recent years have been the shrinking Great Red Spot and/or the Red Spot Hollow, the White Ovals in their various interactions and mergers, and the dark

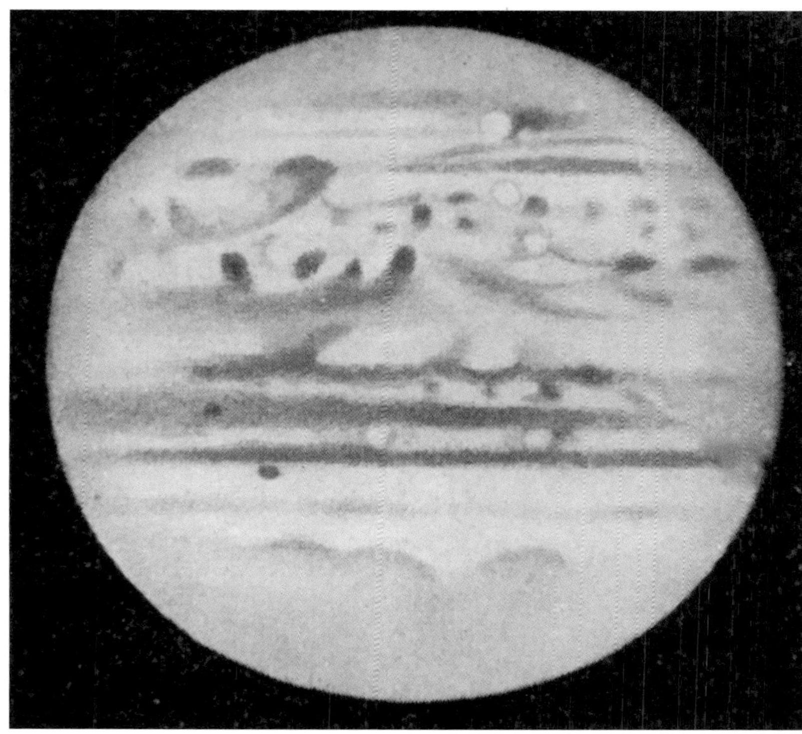

A masterpiece of astronomical art. Jupiter, 5 November 1928. 9h 40m, GMAT as drawn by the great T.E.R. Phillips, with the 46-cm reflector at Headley Rectory. The SEB revival is in full swing now, and the planet is diversified with a plethora of fascinating details.

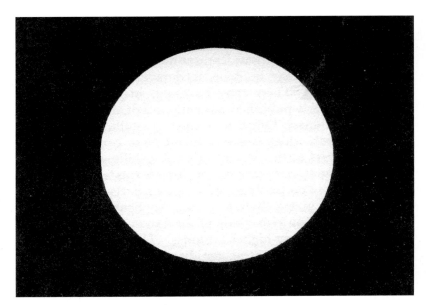

A template for drawing Jupiter, with the polar and equatorial diameters in proportion.

projections of the NEB. Observers especially will want to watch carefully for the major disturbances that occur every few years, for instance the fading of the SEB followed by a dramatic 'revival'. The 'NTB' revivals and the 'broadening' cycles of the NEB, discussed earlier, are also of interest.

It is very useful to measure the longitudes of spots or other features in order to determine which of the currents they belong to. The timings are of course most useful if accompanied by sketches of the features being timed, and as has been the practice since Stanley Williams's time, longitude determinations still involve estimating the transits across the Central Meridian (CM). It is sufficient if the timings are accurate to the nearest minute; this corresponds to only 0.6 degrees of rotation in longitude. The longitude should then be calculated in System I or II. A dedicated observer's logbook is likely to resemble an accountant's, and contain entries like the following, which is a rather typical extract from the logbook kept by the great Jupiter observer Bertrand M. Peek:

30 January 1943

Projection N.T.Bs	$18^h\ 46^m$
f. end of long d. streak in N.E.Bn.	18 47
Little w. spot in N.Temp.Z.	18 47
Grey condensation on Equator	18 47
D.Streak in N. component S.T.B.	18 48
f. end of long d. projection N.E.Bs	18 49
Light spot N. of N.E.B.	18 49

Because of its low surface brightness and marked limb darkening, Jupiter has never been a very satisfactory object for traditional (silver halide) photography. Thus, even during the photographic era, the best work continued to be done with visual methods. However, new technology has produced a revolution. Beginning in the early years of the twenty-first century, amateurs have largely used webcams combined with image-processing software to select and stack the sharpest images. Based on the original webcam technology, planetary cameras have been designed specifically for digital planetary observations, and are able to capture videos containing hundreds of images at a rate of 15 to 200 frames per second, and then, using software such as RegiStax or Autostakkert!, the images with the highest-spatial-frequency components are aligned and stacked before being subject to further image processing. Needless to say, this has become a highly specialized area, and unfortunately discussion of it in further detail lies outside the scope of the present work. A number of reference works are available.[1] There are many amateurs who have distinguished themselves in this kind of work. Among the pioneers of Jupiter CCD imaging, the late Donald C. Parker, of Coral Gables, Florida, deserves mention. Among the many outstanding imagers currently active, Leo Aerts of Belgium and Damian Peach of England are widely regarded as outstanding.

The work available to the well-equipped amateur with a modest telescope and CCD capability is essentially endless. Amateurs

top left: Remotely operated from England, Damian Peach obtained this image of Jupiter on 28 February 2017 using a 1-m Cassegrain near Cerro Pachon, Chile, and an ASI174 camera. This and the following images show different aspects round the planet. Here the Great Red Spot is at the extreme right, and is following by a wake of turbulent white and brown clouds. Note the prominent blue plumes in the Equatorial Zone.

centre: 16 March 2017. The Great Red Spot has moved to the extreme left. Red Spot, Junior, appears just to the left of the central meridian, and below it are a series of White Ovals. Again, note the bluish plumes in the Equatorial Zone, and the serpentine whitish cloud in the broad North Equatorial Belt.

bottom left: 15 March 2017. This shows the opposite hemisphere from that shown in the centre image. The South Equatorial Belt has peculiar structure, with brownish clouds slanting diagonally from the right to the left of the frame; while the North Equatorial Belt is full of intricate white arabesques.

bottom right: 7 March 2017. Note the rich browns in the North Equatorial Belt and in the blue-grey North Polar Region some of the oval storm features that were so prominent in the images sent from the Juno spacecraft.

worldwide are now able to monitor changes in the structure and evolution of atmospheric features on the giant planet, and do so on a daily basis (except, of course, when the planet is on the other side of the Sun and therefore invisible). They provide a continuous record of the large-scale climatic cycles on Jupiter and furnish broad geographical coverage of events (necessary because of the fast rotation of Jupiter). Although Jupiter is regularly monitored by professional astronomers as part of long-term observing programs (and a few observations are made from time to time with the Hubble Space Telescope), these observations are necessarily intermittent. Moreover, most of the professional studies have been carried out in the infrared (IR). Amateurs, on the other hand, are able to image in the red (~580–670 nm), green (~500–580 nm) and blue (~390–500 nm), and so provide multi-wavelength coverage. In addition, more and more amateurs are beginning to image with IR filters (~680–900 nm), which provide slightly deeper penetration into the Jovian clouds; with the narrow-band methane filter (890 nm), which shows light reflected from the highest clouds; and in the near-ultraviolet (less than 390 or 360 nm, depending on the manufacturer).

Since the onset of some of the most interesting Jovian phenomena is unpredictable, amateurs are usually the first to detect them. This is emphatically the case with impact events, all of which, with the exception of the SL9 impacts of 1994, have been discovered by amateurs.

Even with Juno now in orbit round Jupiter, amateurs will continue to play an indispensable role. The majestic planet remains, in the first decades of the twenty-first century, the amateur's planet. Though it is an old planet, having formed soon after the Sun did, 4.6 billion years ago, when it assumed the dominant role in the solar system it has held on to ever since, it is also forever young, and as fresh as every night's observation. It will continue, into the far forseeable future, to reward the student equipped with a small telescope and pursuing visual or CCD observations. As with most things in life, the rewards will be roughly in proportion to the time and effort taken, and to the skills developed with practice and application.

Chances are Jupiter is well placed tonight. Happy observing!

Appendix I

Jupiter by the Numbers

Orbit Radius around the Sun

Metric: 778,340,821 km
English: 483,638,564 miles
Scientific Notation: 7.7834082×10^8 km
Astronomical Units: 5.2028870 AU
By Comparison: 5.203 × Earth

Mean Orbit Velocity

Metric: 47,002 km/h
English: 29,205 mph
Scientific Notation: 1.3056×10^4 m/s
By Comparison: 0.438 × Earth

Orbit Eccentricity

0.04838624
By Comparison: 2.895 × Earth

Equatorial Inclination

3.1 degrees

Equatorial Radius

Metric: 69,911 km
English: 43,440.7 miles

Scientific Notation: 6.9911 × 10^4 km
By Comparison: 10.9733 × Earth

Equatorial Circumference

Metric: 439,263.8 km
English: 272,945.9 miles
Scientific Notation: 4.39264 × 10^5 km
By Comparison: 10.9733 × Earth

Volume

Metric: 1,431,281,810,739,360 km^3
English: 343,382,767,518,322 $miles^3$
Scientific Notation: 1.43128 × 10^{15} km^3
By Comparison: 1321.337 × Earth

Mass

Metric: 1,898,130,000,000,000,000,000,000,000 kg
Scientific Notation: 1.8981 × 10^{27} kg
By Comparison: 317.828 × Earth

Density

Metric: 1.326 g/cm^3
By Comparison: 0.241 × Earth

Surface Area

Metric: 61,418,738,571 km^2
English: 23,713,907,537 sq. miles
Scientific Notation: 6.1419 × 10^{10} km^2
By Comparison: 120.414 × Earth

Surface Gravity

Metric: 24.79 m/s^2
English: 81.3 ft/s^2

By Comparison: If you weigh 100 lb on Earth, you would weigh 253 lb 'on' Jupiter

Escape Velocity

Metric: 216,720 km/h
English: 134,664 mph
Scientific Notation: 6.020×10^4 m/s
By Comparison: 5.380 × Earth

Sidereal Rotation Period

0.41354 Earth Days
9.92496 Hours
By Comparison: 0.41467 × Earth

Effective Temperature

Metric: -148 °C
English: -234 °F
Scientific Notation: 125 K

Atmospheric Constituents

Hydrogen, Helium
Scientific Notation: H_2, He

Number of Satellites

67
Source: NASA

Jupiter data, including images, is archived in NASA's Planetary Data System – Planetary Atmospheres Node, located at New Mexico State University, Las Cruces, New Mexico: http://pds-atmospheres.nmsu.edu

APPENDIX II

CONJUNCTIONS OF JUPITER

- Jupiter and Saturn will be very close to each other in the sky on:
  - 21 December 2020
  - 31 October 2040
  - 7 April 2060

The last such conjunction happened in May 2000.

- Jupiter and Venus will be very close to each other in the sky on:
  - 31 October 2040
  - 7 April 2060
  - 15 March 2080

The last such conjunction happened in December 2000.

- Jupiter and Mars will be very close to each other in the sky on:
  - 7 January 2018
  - 20 March 2020
  - 29 May 2022
  - 14 August 2024
  - 16 November 2026
  - 19 July 2029
  - 28 September 2031
  - 1 December 2033
  - 18 February 2036
  - 22 May 2038

18 August 2040
30 October 2042
4 January 2045
18 March 2047
11 June 2049
8 September 2051
24 November 2053
1 February 2056
12 April 2058
1 July 2060

The last such conjunction happened in October 2015.

Based on tables by Richard Nolle.

APPENDIX III

THE GALILEAN SATELLITES BY THE NUMBERS

	Mass (10^{20} kg)	Radius (km)	Mean density (kg/m^3)	albedo
Io	893.2	1821.6	3530	0.62
Europa	480.0	1560.8	3010	0.68
Ganymede	1481.9	2631.2	1940	0.44
Callisto	1075.9	2410.3	1830	0.19

Source: NASA

The rest of Jupiter's satellites are at least an order of magnitude smaller. Only one, Amalthea, was discovered by eye with a telescope. Not massive enough for their gravity to compress them into the least volume, these satellites are aspherical. Certain minor satellites have similar orbital properties and are thought to be the fragments left by a collision of slightly larger bodies.

Appendix IV

Space Probes to Jupiter

Pioneer 10

Mission: first flyby of Jupiter

Launched	3 March 1972
Arrived Jupiter	4 December 1973

Pioneer 11

Mission: first polar views of Jupiter on way to fly by Saturn

Launched	6 April 1973
Arrived Jupiter	2 December 1974

Voyager 1

Mission: flyby of Jupiter and Saturn

Launched	5 September 1977
Arrived Jupiter	5 March 1979

Voyager 2

Mission: flyby of Jupiter en route to Saturn, Uranus and Neptune

Launched	20 August 1972
Arrived Jupiter	9 July 1979

Galileo

Mission: first to orbit Jupiter. Dropped secondary probe into Jupiter's atmosphere

Launched	18 October 1989
Arrived Jupiter	7 December 1995

Cassini

Mission: flyby of Jupiter en route to orbit Saturn. Dropped probe onto Saturn's satellite Titan

Launched	15 October 1997
Arrived Jupiter	30 December 2000

New Horizons

Mission: flyby of Jupiter en route to Pluto

Launched	19 January 2006
Arrived Jupiter	28 February 2007

Juno

Mission: first Jupiter polar orbiter

Launched	5 August
Arrived Jupiter	4 July 2016

Source: NASA

Glossary

Aphelion: The point in a planet's orbit at which it is farthest from the sun.

Apparition: The period of time during which an object such as a planet is visible and not obscured by the glare of the Sun.

CCD: Charged-couple device. Used for digital imaging.

Coronograph: A telescope that blocks out the bright layers of the Sun, allowing observation of dimmer objects in nearly the same angular direction as the Sun.

Kuiper Belt: A band of small icy bodies that orbits the Sun in the ecliptic plane past Neptune. Pluto is an example of a Kuiper Belt object.

Lagrangian Point: A stable orbit about a body, such as the Sun or a planet, due to the balance of the gravitational force from two other solar-system bodies.

Oort Cloud: A theoretical shell of potential comet nuclei far from the Sun that is a source of long-period comets.

Perihelion: The point in a planet's orbit at which it is closest to the Sun.

Planetesimal: A small, primitive body in the early solar system, a multitude of which collided with each other to form planets.

Resonance: The enhanced gravitational effect at a particular place in a revolving body's orbit, due to the synchronous orbit of another, nearby body.

References

Prologue

1 Noel Swerdlow, *Babylonian Theory of the Planets* (Princeton, NJ, 1998), p. 94.

1 The Jovian Planets

1 Percival Lowell, *The Evolution of Worlds* (New York, 1909), p. 14.
2 L. Billings, 'In Search of Alien Jupiters', *Scientific American*, CCCXIII/2 (August 2015), p. 40.
3 A. Boss, 'Orbital Migration of Protoplanets in a Marginally Gravitationally Unstable Disc', *Astrophysical Journal*, DCCLXIV (2015), p. 194.
4 H. Levison, R. Kretke and M. Duncan, 'Growing the Gas-giant Planets by the Gradual Accumulation of Pebbles', *Nature*, DXXIV (20 August 2015), p. 322.
5 Her seminal papers are R. Malhotra, 'The Origin of Pluto's Peculiar Orbit', *Nature*, CCCLXV (1993a), p. 819, and 'Orbital Resonances in the Solar Nebula: Strengths and Weakness', *Icarus*, CVI (1993b), p. 254.
6 R. Malhotra and J. G. Williams, 'Pluto's Heliocentric Orbit', in *Pluto and Charon*, ed. S. A. Stern and D. J. Tholen (Tucson, AZ, 1997), p. 148.

3 Superficial Matters

1 T. W. Webb, *Celestial Objects*, 6th edn (London, 1917), p. 190.
2 Richard A. Proctor, *Other Worlds than Ours: The Plurality of Worlds Studied in Light of Recent Science* (New York, 1896), p. 152.
3 Percival Lowell, *The Evolution of Worlds* (New York, 1909), p. 167.

4 Atmospherics

1 Christiaan Huygens, *Cosmotheoros* (London, 1698).

5 The Great Red Spot Becomes Great

1 John H. Rogers, *The Giant Planet Jupiter* (Cambridge, 1995), p. 6.
2 J. H. Schröter, *Beobachtungen Verschiedener Schwarzdunkler Kleiner Flecken des Jupiters, Welche von Sehr Kurzer Dauer im Verhältnis mit der von Cassini Bestimmten Umdrehungszeit des Jupiters von Einer Merklich Geschwindern Bewegung Erschienen* (Lilienthal, 1786).
3 Thomas Hockey, *Galileo's Planet: Observing Jupiter Before Photography* (Bristol, 1999), p. 140.
4 C. Pritchett, 'Markings on Jupiter', *Observatory*, II (1879), p. 307. See also Hockey, *Galileo's Planet*, p. 141.
5 E. E. Barnard, 'A Few Unscientific Experiences of an Astronomer', *Vanderbilt University Quarterly*, VIII (1908), pp. 273, 281. The most thorough biography of Barnard is William Sheehan, *The Immortal Fire Within: The Life and Work of Edward Emerson Barnard* (Cambridge, 1995).
6 Percival Lowell, 'Photographs of Jupiter', *Philosophical Magazine and Journal of Science*, 6th series, XIX (1910), p. 92.
7 Percival Lowell, *The Evolution of Worlds* (New York, 1909), p. 164.
8 Cited in Rogers, *The Giant Planet Jupiter*, p. 13.
9 Stanley Williams, *Zenographical Fragments: The Motions and Changes of the Markings on Jupiter in 1888*, vol. II (London, 1909), p. 9.
10 Ibid., p. 12.
11 Stanley Williams, 'On the Drift of the Surface Material in Different Latitudes', *Monthly Notices of the Royal Astronomical Society*, LVI (1896), p. 143.
12 Richard Baum, personal communication to W. S.
13 Williams, 'On the Drift of the Surface Material in Different Latitudes', p. 143.
14 See R. McKim, 'P. B. Molesworth's Discovery of the Great South Tropical Disturbance on Jupiter, 1901', *Journal of the British Astronomical Association*, CVII/5 (1997), p. 239.
15 Bertrand Peek, 'The First Fifty Years of the British Astronomical Association', *Memoirs of the British Astronomical Association*, XXXVI/2 (1948), p. 98.
16 Rogers, *The Giant Planet Jupiter*, p. 43.
17 Ibid., p. 45.

6 A Bewildering Phantasmagoria: Jovian Meteorology

1 Personal communication to W. S.
2 As noted in E. Kardasi et al., 'The Need for Professional–amateur Collaboration in Studies of Jupiter and Saturn', *Journal of the British Astronomical Association*, CXXVI/1 (2016), p. 29.

3 Quoted in Bertrand M. Peek, *The Planet Jupiter* (London, 1958), p. 159.
4 Thomas Dobbins and William Sheehan, 'Jupiter's Deep Mystery', *Sky and Telescope* (December 1999), p. 118.
5 Bertrand Peek, *The Planet Jupiter* (London, 1958), p. 183.
6 A. Sanchez-Lavega and J. M. Gomez, 'The South Equatorial Belt of Jupiter, I: Its Life Cycle', *Icarus*, CXXI (1996), pp. 1–17, and A. Sanchez-Lavega, et al., 'Depth of a Strong Jovian Jet from a Planetary-scale Disturbance Driven by Storms', *Nature*, CDLI (2008), pp. 437–40.
7 Rory Barnes and Thomas Quinn, 'A Statistical Examination of the Short-term stability of the U Andromedae Planetary System', *Astrophysical Journal*, DL/2 (1 April 2001), p. 884.

7 Above Jupiter

1 See Jay M. Pasachoff, 'Simon Marius's *Mundus Iovialis*: 400th Anniversary in Galileo's Shadow', *Journal for the History of Astronomy*, XLVI/2 (May 2015), p. 218.
2 Galileo Galilei, *Sidereus nuncius or the Sidereal Messenger*, trans. with introduction, conclusion and notes by Albert Van Helden (Chicago, IL, 1989), p. 84.
3 S. J. Peale, P. Cassen and R. T. Reynolds, 'Melting of Io by Tidal Dissipation', *Science*, CCIII (1979), p. 892.

8 Jupiter in Collision

1 Thomas Hockey, 'The Shoemaker-Levy 9 Spots on Jupiter: Their Place in History', *Earth, Moon, and Planets*, LXVI/1 (1994), p. 1.
2 Thomas Hockey, *Galileo's Planet: Observing Jupiter Before Photography* (Philadelphia, PA, 1999), p. 143.

9 Juno to Jupiter

1 The preliminary results from Juno have been published in S. J. Bolton et al., 'Jupiter's Interior and Deep Atmosphere: The Initial Pole-to-pole Passes with the Juno Spacecraft', *Science*, CCCLVI/6340 (26 May 2017), pp. 821–5; and J.E.P. Connerney et al., 'Jupiter's Magnetosphere and Aurorae Observed by the Juno Spacecraft during its First Polar Orbits', *Science*, CCCLVI/6340 (26 May 2017), pp. 826–31.

10 Observing Jupiter

1 The authors have found the following useful: (1) Now a bit dated but still a useful overview: Steve Massey, Thomas A. Dobbins and Eric J. Douglass, *Video Astronomy* (Cambridge, MA, 2004); (2) John W. McAnally, *Jupiter and How to Observe It*, Observer's Guide Series (New York, 2008); (3) Emmanuel Kardasis et al., 'The Need for Professional–amateur Collaboration in Studies of Jupiter and Saturn', *Journal of the British Astronomical Association*, CXXVI/1 (2016), p. 229.

Acknowledgements

Jupiter, with its large disc and copious, ever-changing detail, has always been one of the most rewarding objects in the heavens for amateur observers, and amateurs – many belonging to the British Astronomical Association and the Association of Lunar and Planetary Observers – have contributed immensely to an understanding of the planet's circulation and characteristic cloud features. It is pleasant to record that some of the greatest of them – Stanley Williams, who defined the nine main currents, and Elmer Reese, who discovered the internal rotation of the Great Red Spot – did some of their important work with reflectors of only 15 cm aperture.

Even in the modern era, in which spacecraft have been placed into orbit around the planet – including the polar-orbiting Juno spacecraft, which arrived in June 2016 – amateurs continue to play an important role. The monitoring of the Giant Planet's meteorology with small telescopes provides an overview that allows comparison with the centuries' long baseline of ground-based observations, and, in addition, phenomena such as impacts of comets into the Jovian atmosphere are short-lived and apt to be missed by all except dedicated amateurs keeping video-imaging vigils over long periods of time. Since the famous SL9 impacts of July 1994, four impacts so far have been recorded by amateur video imagers, and it now appears that such events are not uncommon.

Both of the authors are long-standing students of the Giant Planet. Co-author William Sheehan made his first drawings of the planet with small telescopes in the 1960s, and has continued to observe and sketch the planet ever since; while Thomas Hockey has written *Galileo's Planet*, which is still a definitive source on eighteenth- and nineteenth-century observations of Jupiter. The authors express their gratitude to their mentors in the study of the Giant Planet, who, in Hockey's case, was Reta Beebe of New Mexico State University, his dissertation advisor, and in Sheehan's case was John H. Rogers, long-time Director of the Jupiter Section of the British Astronomical Association. In addition, several authorities commented on early drafts of the manuscript, including John Rogers of the BAA, the late Peter Hingley of the RAS Library, and Amy Simon and Robert Suggs of NASA.

Lauren Amundson of the Lowell Observatory provided illustrations and granted permission to reproduce them. Above all, we would like to thank Peter Morris for believing in the project, and Reaktion Books for publishing it.

Photo Acknowledgements

The author and publishers wish to express their thanks to the below sources of illustrative material and/or permission to reproduce it:

© Leo Aerts: pp. 14, 72 (bottom), 122 (top and bottom), 123; © Julian Baum: p. 9; from E.E. Barnard, *Milky Way and Comet Photographs, Publications of the Lick Observatory*, vol. XI (1913): p. 20; photos Serge Brunier/European Southern Observatory: pp. 22–3; photos British Astronomical Association: pp. 85 (bottom), 86 (bottom), 87 (top and bottom), 103 (column 1, bottom), 103 (column 2, top and middle); from *Memoirs of the British Astronomical Association*, vol. XXX (1935): pp. 103 (column 1, middle), 163; from George F. Chambers, *Astronomy* (London, 1913): p. 72 (top); from Agnes Clarke, *A Popular History of Astronomy during the Nineteenth Century* (London, 1908): p. 74 (left); collection of William Sheehan: pp. 17, 54, 68, 71 (bottom right), 73 (right), 121, 144–5, 164; from A. Favoro, ed., *Le opere di Galileo Galilei* (Florence, 1890–1909): p. 52; from Camille Flammarion, *Les Terres du Ciel: voyage astronomique sur les autres mondes et description des conditions actuelle de la vie sur les diverse planètes du système solaire* (Paris, 1884): pp. 42, 58, 119; courtesy Peter Hingley, the Royal Astronomical Society, London: pp. 65 (top), 66 (bottom), 67 (top and bottom), 71 (top and bottom left); photo Hubble Space Telescope/NASA/Goddard Space Flight Center [GSFC]: p. 48; from Christiaan Huygens, *Œuvres Complètes de Christiaan Huygens*, vol. XV (La Haye, 1925): p. 55; photo Musée de Louvre: p. 10; photo courtesy Lowell Observatory: p. 39; from Percival Lowell, *The Evolution of Worlds* (New York, 1909): p. 81; National Maritime Museum, Greenwich, London (Caird Collection): p. 53; photo National Library of Medicine: p. 62; photos NASA: pp. 75, 114, 159 (top); photo NASA/ESO: p. 124; photo NASA/Johns Hopkins University Applied Physics Laboratory: p. 135; photo NASA/Johns Hopkins University Applied Physics Laboratory [JHUAPL]/Southwest Research Institute [SWRI]: pp. 120, 138; photo NASA/Hubble Space Telescope: p. 150; photo NASA/Hubble Space Telescope and Jupiter Imaging Team: p. 147; photos NASA/JPL: pp. 97 (left), 98, 107 (top and bottom), 125, 128, 146; photo NASA/JPL/Ames

Research Center: p. 131; photos NASA/JPL-Caltech/DLR: pp. 130 (top), 133, 144; photos NASA/JPL-Caltech/SwRI/MSSS: pp. 154, 158 (courtesy Roman Tkachenko); photos NASA/JPL-Caltech/SwRI/MSSS/Gerald Eichstädt and John Rogers: pp. 99, 160; photo NASA/JPL-Caltech/SwRI/MSSS/Jason Major: p. 74 (bottom); photo NASA/JPL/Cornell University: p. 117; photo NASA/JPL/University of Arizona: p. 130 (bottom); photo NASA/Space Science Institute: p. 97 (right); photo NASA/H. Weaver (Johns Hopkins University and T. Smith/Space Telescope Science Institute): p. 143; The New York Public Library: p. 70; © Damian Peach: pp. 6, 166 (all); from T.E.R. Phillips, ed., *Hutchinson's Splendour of the Heavens* (London, 1925): pp. 28, 73 (left), 77, 84, 85 (top); from *The Philosophical Transactions of the Royal Society*, vol. CXVIII (London, 1858): p. 49; from *Publications of the Astronomical Society of the Pacific*, vol. I (1889): p. 69; courtesy John Rogers, Director of the Jupiter Section of the British Astronomical Association: pp. 57, 103 (column 1, top, and column 2, bottom); courtesy of the Royal Astronomical Society: p. 66 (top); photos William Sheehan: pp. 13, 19, 64, 86 (top), 104; © Brad Smith: p. 18; Solar Dynamics Observatory/HMI image: p. 65 (bottom); photo Vatican Museums (Pinacoteca Vaticana): p. 60; from Arthur Stanley Williams, *Zenographical Fragments: the Motions and Changes of the Markings on Jupiter in 1888* (London, 1909): p. 80.

Index

Page numbers in ***bold italics*** refer to illustrations